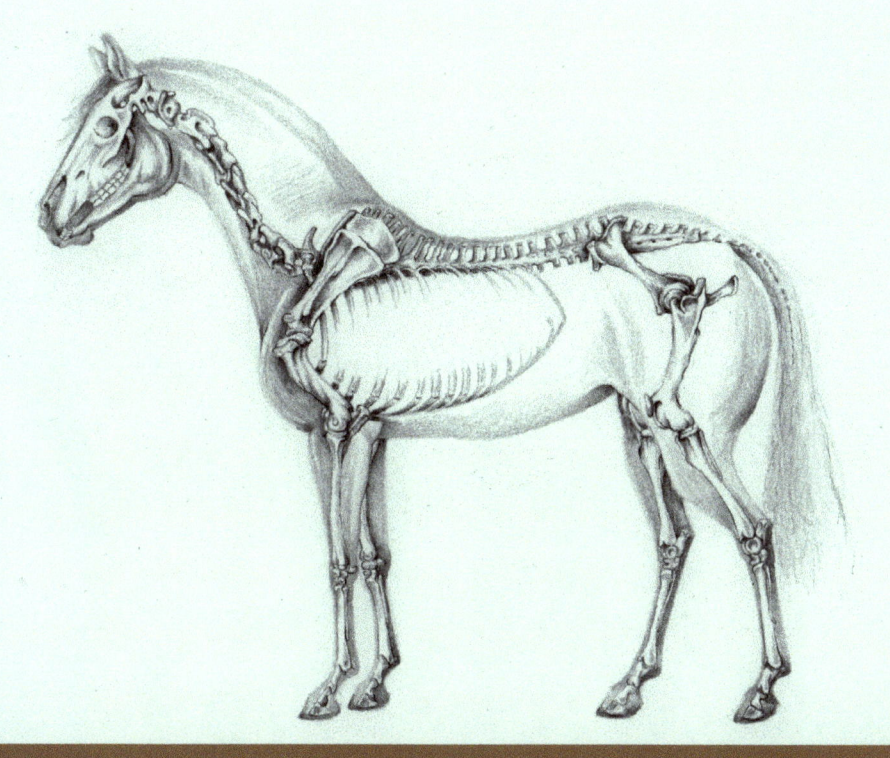

A Journey through the Horse's Body

Dr. Christina Fritz

A Journey through the Horse's Body

I would like to thank Lars and Lennard Wittenburg for their patience and support in the creation of this manuscript during the family holiday.

Current Edition:
Copyright © 2022 Xenophon Press
ISBN: 978-1948717427
Design: Ravenstein, Verden
Translation: Helen McKinnon
Editorial of the English edition: Christopher Long
Cover image: Edition Boiselle/Susanne Retsch-Amschler
Photographs inside book: Cadmos archive, Dr. Christina Fritz, Kai Kreling, Dr. Richard Maurer, Christiane Slawik
Illustrations: Juliane Denmann, Maria Mähler, Susanne Retsch-Amschler
All rights reserved: No part of this book may be reprinted or reproduced or utilized in any form or by any electronic, mechanical, or other means, now known or hereafter invented, including photocopying and recording, or in any information storage or retrieval system, without permission in writing from the publisher.

Published by Xenophon Press LLC
Franktown Virginia USA 23354
XenophonPress@gmail.com
1-757-442-1060

Previous Edition:
ISBN: 978-0857880062
Copyright ©2012 Cadmos Publishing Limited, Richmond, UK
Copyright of original edition © 2011 Cadmos Verlag GmbH, Schwarzenbek, Germany

British Library Cataloguing in Publication Data
A catalogue record of this book is available from the British Library

Contents

Foreword 6

Basic Anatomical Principles 7
    Riding and Veterinary Latin 8

The Musculoskeletal System 12
    The Bones 13
    The Structure of Different Bones ... 14
    The Joints 15
    The Skeleton 22
        The Spinal Column 22
        The Skull 26
        The Teeth 27
        The Limbs 30
        Structure of the Distal Limb
        and Hoof 38
    The Musculature 42
        Muscle Types and
        Muscle Function 42
        Superficial, Medial and Deep
        Musculature of the Trunk 44
        The Musculature of the Forehand ... 48
        The Musculature of
        the Hindquarters 51

The Cardiovascular System 56
    The Blood Cells 57
    The Vascular System 58
    The Structure of the Heart 61

The Lymphatic System 64

The Respiratory System 66

The Digestive System 69
    The Stomach 70
    The Small Intestine 71
    The Large Intestine 72
    The Liver 74
    The Pancreas 74

The Genitourinary System 75
    Kidneys and Bladder 75
    Sexual Organs and Reproduction ... 78

The Skin and Skin Appendages 83

The Nervous System 86
    The Central Nervous System 87
    The Autonomic Nervous System ... 89
    The Peripheral Nervous System ... 90
    The Horse's Senses 90

Index 94

Foreword

Dear Reader,

Welcome to a voyage of discovery through the horse's body. In this book, I will describe to you the structure and function of all of the parts of the horse's body. The book features extensive image material to help clarify the theoretical explanations. By understanding anatomy, you will be able to ride your horse better and keep him in an environment that suits him best.

Look at your horse in a whole new light. See him not just as your friend and partner, but as the steppe animal, the herd animal, the flight animal that has evolved over millennia to become better and better adapted to his environment. Above all, the horse is an animal of movement that, in the wild, would be active for up to 16 hours a day in search of food. This is only possible with a specially adapted musculoskeletal system, to which a particularly detailed chapter of this book has been dedicated. The cardiovascular system and the respiratory system are also adapted to this life in motion. The horse's digestive system is perfectly suited to the diet of a steppe animal and is fundamentally different from that of the carnivorous dog or omnivorous human. I will also go into these special features in greater depth in this book.

Embark on a journey through the horse's body and you will never look at your horse in the same way again!

Basic Anatomical Principles

People have been interested in anatomy since the dawn of time. The term "anatomy" is derived from the Latin *anatemnein*, which means "dissect" or "dismember". Correspondingly, we owe most of our anatomical knowledge to the people who dissect the dead and make the functional units visible. By breaking down the body into its smallest units, we can understand its functions and connections and therefore the body as a whole. This means that anatomy is not only the basis for veterinary treatment, i.e. understanding which body part no longer works so that it can be fixed, but also the basis for correct riding.

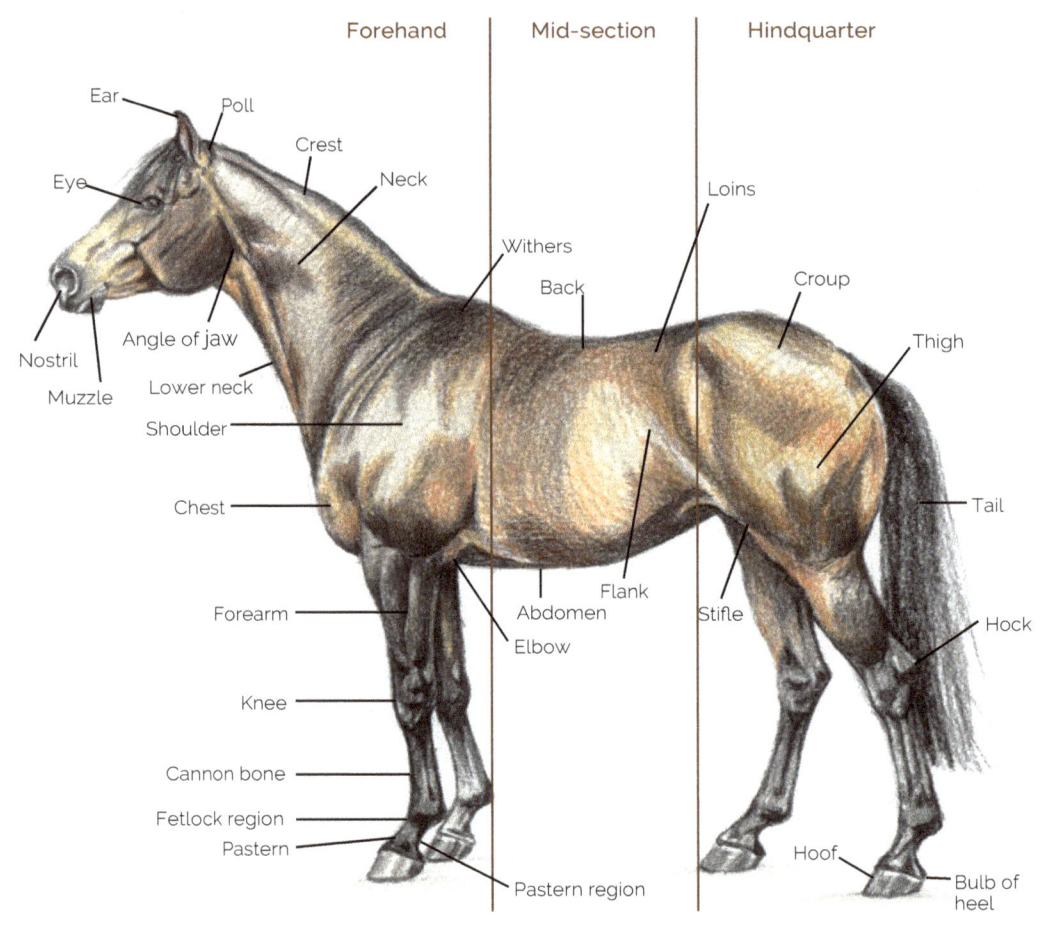

The points of the horse. (Illustration: Retsch-Amschler)

Riding and Veterinary Latin

You will come across many Latin terms in this book, because Latin has always been the language of medicine. But don't worry; it's not as difficult as it sounds!

Some of the Latin names have already found their way into our language. For example, many riders will be familiar with the *longissimus*, the long back muscle that transfers the power of the hindquarters to the forehand, like an elastic band. Many of these terms are very flowery, such as *os sacrale*, the sacrum, whose name is derived from its cross shape. Others have arisen from comparative anatomy between the species. A *musculus extensor* (literally translated: extensor muscle) may also have a bending function, depending on how the joints are positioned in relation to one another in the animal concerned. Not all animals are unguligrades like horses, which walk on the tips of their toes like ballerinas. Many are digitigrades, such as cats or dogs, for example and some animals are plantigrades, such as human beings or badgers. Nevertheless, the same muscle always has the Nevertheless, the same muscle always has the same name in all animals.

The Latin terms in this book will also enable you to understand your vet or equine osteopath better when they are discussing your horse's health problems with you. We will also use non-Latin words in this book that have made their way into our language, such as knee joint or fetlock. Terms have been adopted into the language of riding that are not anatomically correct, such as "knee" for the carpal joint, which can lead to misunderstandings.

Terms that come both from the language of riding and the language of medicine are used to describe anatomy. In anatomy, the obvious English terms such as "above" or "below" are not used to describe directions, because these

The adduction/abduction movement can easily be recognised in the half-pass. (Photo: Slawik)

terms can change, for example, if the horse has to be turned onto its back for an operation. That is why we use Latin terms such as *dorsal* or *ventral*. In equitation we also use words to describe direction of movement. Adduction describes pulling the leg towards the midline and abduction describes the lateral stretching of the leg away from the midline. The adduction/abduction movement is particularly important for the half-pass.

Flexion, i.e. upwards curving, of the back is the aim of all riding. In order for the back to flex, the sacrum, i.e. the end

Basic Anatomical Principles 9

We can clearly see how horses arch their backs when they buck. (Photo: Slawik)

This horse's back is clearly extended. (Photo: Slawik)

of the spinal column, and the pelvis, must tilt into a "sitting" position. This movement is also described as flexion. The opposite movement, i.e. when the back is pushed down, is called extension, as is the movement of the plevis and sacrum when stretching out the hindquarters.

An Overview of Anatomical Terms

Regions:
Forehand: All of the parts of the horse from the nose to the shoulder, including both front legs.
Midsection: The area between the forehand and the hindquarters, i.e. the back, chest and abdominal cavity with their respective organs.
Hindquarters: Everything behind the flank, i.e. croup and back legs.

Parts of the body:
Equine body parts have names very similar to human body parts: head, neck, shoulder, upper arm, forearm, back, abdomen, chest, hip and thigh. The only difference is that horses walk on four legs, i.e. on their hands and feet or, more precisely, on the tips of their toes. Everything below the knee joint or hock is called the "distal limb".

Terms of location:
Cranial: Forwards or in the direction of the head. With regard to the head, we also say rostral for the direction of the nose or caudal for the direction of the ears.
Caudal: Backwards or towards the tail.
Dorsal: Upwards or towards the back.
Ventral: Downwards or towards the abdomen.
Medial: Towards the midline of the body. In the leg, this means the adduction movement.
Lateral: Sideways away from the midline of the body. In the leg, this means the abduction movement.

Anatomical Terms of Location. (Photo: Slawik)

Terms of location:
Proximal: In the direction of the body.
Distal: In the direction of the foot.
Palmar: The back of the foreleg, i.e. the "palm" side.
Plantar: The back of the hind leg, i.e. the "sole" side.
The fronts of the legs, i.e. the back of the hand or the back of the foot, are described as "dorsal".

Horses are built for movement. (Photo: Slawik)

The Musculoskeletal System

The musculoskeletal system of a horse consists of a passive part, by which we understand the bones and joints and an active part that encompasses all of the muscles. Both parts need to interact in order for the horse to move. Muscles can still contract without bones but it will get them nowhere and a skeleton without muscles could not even stand, because even that requires muscular tension. The skeleton also provides the required stability and the points of influence for the muscles that then move the body.

The Bones

The role of the bones in the musculoskeletal system is to act as a support and point of influence for the muscles, tendons and ligaments and to support the weight of the body.

Bone is not dead tissue, as you might think from anatomical specimens. Only the mineral calcium component of the bone is left after preparation as a specimen, in the same way that a coral on land is just the dead calcium skeleton of the living organism. The mineral component of bones gives them their strength, which makes bone the hardest substance in the body, after tooth enamel. However, the mineral component, mainly calcium, phosphorous and magnesium, only makes up around half of bone. In an adult horse, the skeleton contains around 9 to 10 kilograms of calcium, around 20 percent of which can be mobilised if required. The bones are therefore one of the body's most important mineral stores. A quarter of the weight of bone is water and another quarter is living bone cells. These bone cells (osteocytes) ensure that a bone can always be grown, broken down or reconstructed and therefore adapted to the strain it is under. The activity of these bone cells is particularly impressive after a bone has been broken, when the gap between the two fragments is closed by new bone material within a few weeks. However, osteocytes are also responsible for the formation of exostoses because they build up extra material to stabilise the bone if the periosteum becomes inflamed or if the slightest crack appears in the bone.

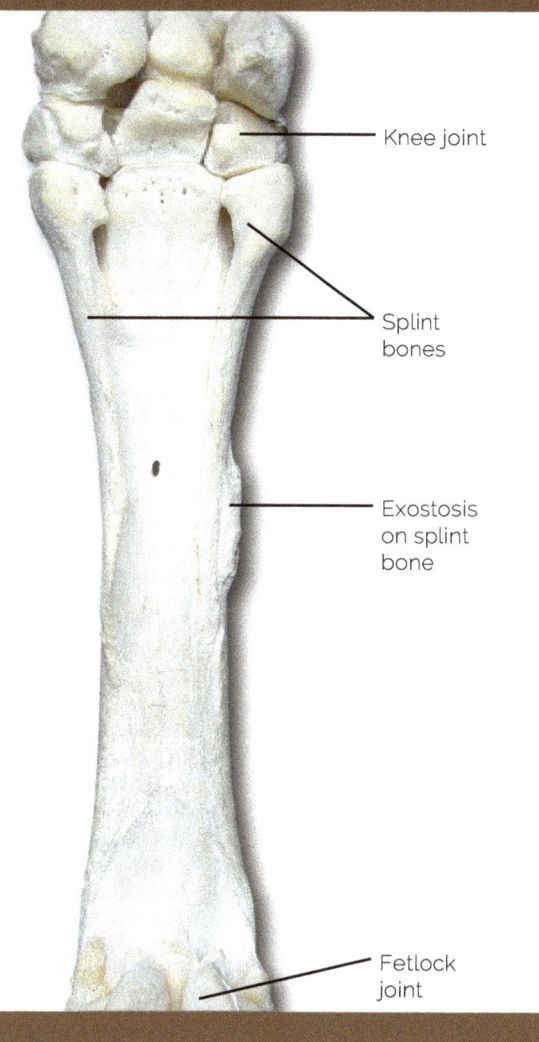

Exostosis on the splint bone, palmar view. (Photo: Fritz)

Did you know ...?

A great deal of weight rests on the cannon bones, which increases further when the horse is ridden. The cannon bones of an unridden horse have a round cross section, which becomes a transverse oval when the horse is ridden. In this instance, the body produces more bone mass to be able to bear more weight.

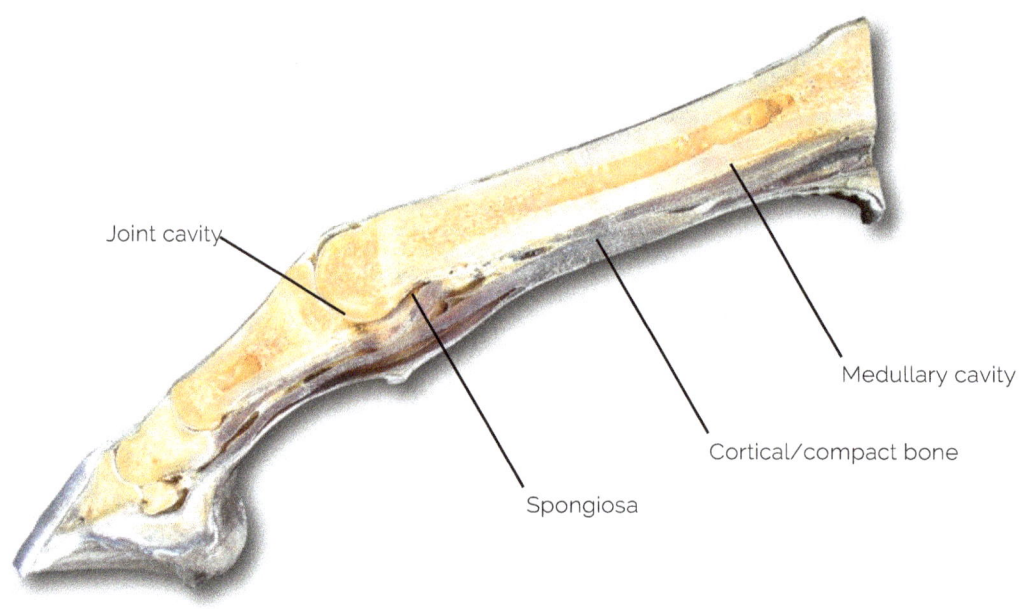

Cross section of a horse's leg. (Photo: Fritz)

The Structure of Different Bones

In terms of structure, we can differentiate between short bones, flat bones and long bones. Short and flat bones are only made from spongy bone material (spongiosa) and the thin, hard cortical substance (corticalis) that covers them. Flat bones include the cranial or pelvic bones, but the ribs are also considered to be flat bones because of their structure. Short bones include, for example, the short pastern bone at the coronet band. All long bones are tubular bones, for example the large, long bones of the limbs. Long bones have a medullary cavity at their centre, which is surrounded by a thick, particularly stable layer of bone (cortical or compact bone). The ends of long bones are made of spongiosa.

Bones have a trabecular structure to achieve the required strength at a simultaneously low weight. That means that they are not only made of a compact mass of calcium as we would initially think from their appearance, but from finely branching trabeculae that are particularly visible in the area of the spongiosa. These trabeculae are aligned along the tensile and compressive forces that act on the bones. This enables the bone to be solid and flexible at the same time.

The periosteum is on the outside of all bones. It is made of firm connective tissue (fibrosa). There is a layer on the inside where the bone cells are, called the osteoblastic zone. Exostoses may be formed here if the periosteum is damaged, by a blow for example. A great many bundles of collagen fibres, known as Sharpey's fibres, extend from the periosteum into the bones and attach them firmly to the periosteum. Blood vessels also penetrate into the bone from the periosteum and supply the osteocytes that lie deep inside the bone, as well as the cells in the bone marrow. At the epiphysis or end part of the bone, the bundles of collagen fibres continue into the upper layer of the articular cartilage and radiate into the fibrous joint capsule. There they ensure a close connection between the articular cartilage and the bone, as well as between the joint capsule and the bone. The tendons and points of influence of the muscles also originate from the periosteum and their fibres go into the bones. Bones and tendons therefore grow firmly together.

The medullary cavity is found inside the long bones and is filled with bone marrow that is well supplied with blood (red bone marrow). Among other matter, the red bone marrow is made up of stem cells for red and white blood cells. Flat bones also contain red bone marrow. While the bone marrow in the long bones reduces in size and only consists of fat (white bone marrow) in older animals, the red bone marrow in flat bones is there for life and forms the blood cells that are essential for existence.

In the embryo, all of the bones start out as cartilage and ossify later because of mineral deposits. In most of the long bones in the skeleton, we can differentiate between a shaft and two end parts. While the animal is still growing, the ends of the bones are still separated from the shaft by a cartilaginous epiphyseal disc (usually called "epiphysis" or "epiphyseal cartilage"). Lengthways growth takes place at the epiphyseal cartilage until the animal is fully grown. Horses are fully grown at around six years, but some do not reach physical maturity until they are eight years old. Hormones, and not just genetics, determine how big a horse will end up being. Therefore geldings are, on average, larger than stallions and grow for longer. Furthermore, diet and the biomechanical forces that act on the skeleton also influence growth. If the horse is put under a lot of strain early on, for example for showjumping or reining, the epiphyseal cartilage ossifies sooner in order to stabilise the bones. The horse will therefore stop growing. In these cases, "splints" can often be felt on the insides of the front legs below the knee joint. Splints are bony proliferations that are formed because of high mechanical strain in this area, in order to prevent the bone from breaking at the epiphyseal cartilage. Once the epiphyseal cartilage has been completely ossified with minerals, growth is no longer possible. Stabilisation and repair processes may take place afterwards, but the length of the bone stays the same.

The Joints

The bones of the skeleton are connected together with joints so that they can move. Different types of joints have developed according to their role and, depending on their function, joints may be highly mobile. These joints are usually covered with muscles, such as the joints in the limbs. There are also joints that can only

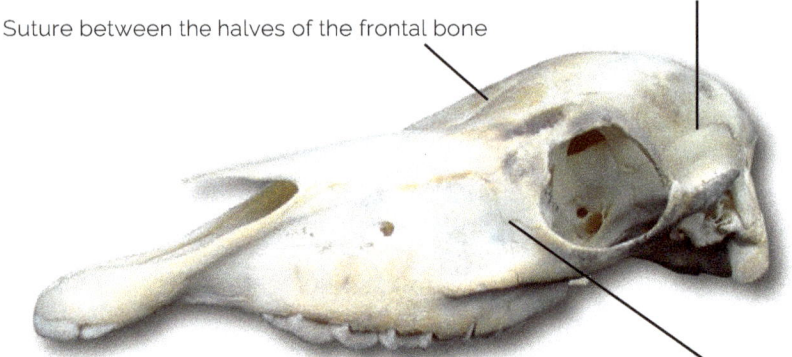

Foal skull with sutures clearly visible. (Photo: Fritz)

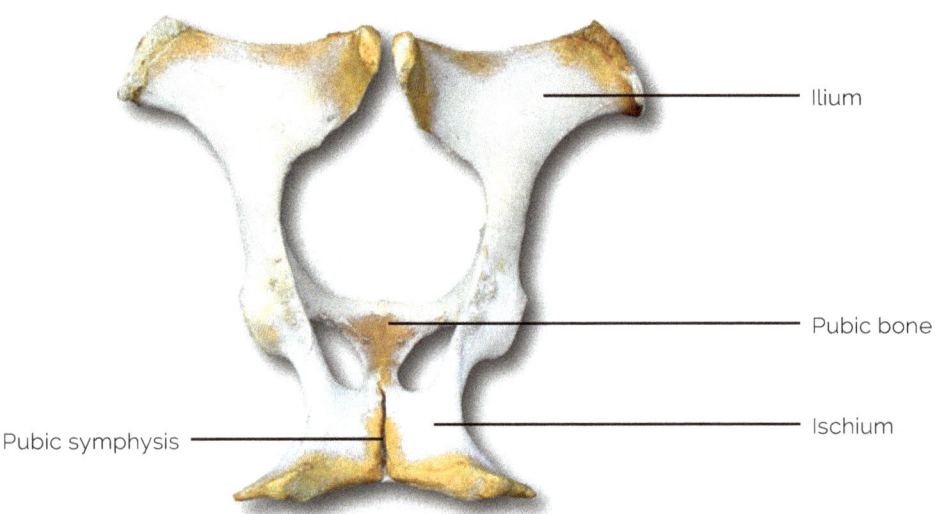

Pelvis with pubic symphysis clearly visible, dorsal view. (Photo: Fritz)

be moved passively. They create scope for movement without actively moving themselves, for example the pubic symphysis, which is the point at which the pubic bones of the pelvis come together. This symphysis, which is normally closed with cartilage, must come apart when giving birth to make the birth canal wide enough for the foal. The skull also consists of individual bones that are connected together so that they can move passively. In these cases, we talk about sutures, which are fine joints that hold the bones together with connective tissue.

Real joints, i.e. joints that are involved in the active movement of the body, all have a similar structure. Joints consist of the ends of the bones that form the joint. They are covered with hyaline articular cartilage. The term "hyaline" comes from the Greek word, meaning "milky", which is exactly what healthy articular cartilage looks like: milky and slightly bluish.

Did you know ...?

Articular cartilage has an important buffer function that cushions the bone ends against one another. To this end, the cartilage cells become filled with more fluid during movement, forming a thick layer of small cushions that absorb impact. Standing for long periods squeezes the fluid out of the cartilage cells. In horses, around 15 to 20 minutes of "warm up" time, i.e. quiet movement at walk, is required for the cartilage cells to be able to properly perform their shock-absorbing task.

Articular cartilage is firmly attached to the bone underneath. The joint cavity is between the two layers of cartilage. Only bone shows up on X-rays, so the joint cavity sometimes looks very wide, because the layers of cartilage, which are often several millimetres thick, are not visible. The joint cavity itself is only very narrow and is filled with clear to pale yellow, slightly slimy fluid (synovial fluid). The role of synovial fluid is to allow the layers of cartilage to glide during movement. It is also the nutritive fluid for the cartilage cells because they are not directly supplied by the vascular system. Synovial fluid is produced by the inner layer of the joint capsule and reaches the cartilage cells by diffusion through the joint cavity. Lots of quiet movement increases diffusion of synovial fluid and improves supply to the cartilage cells.

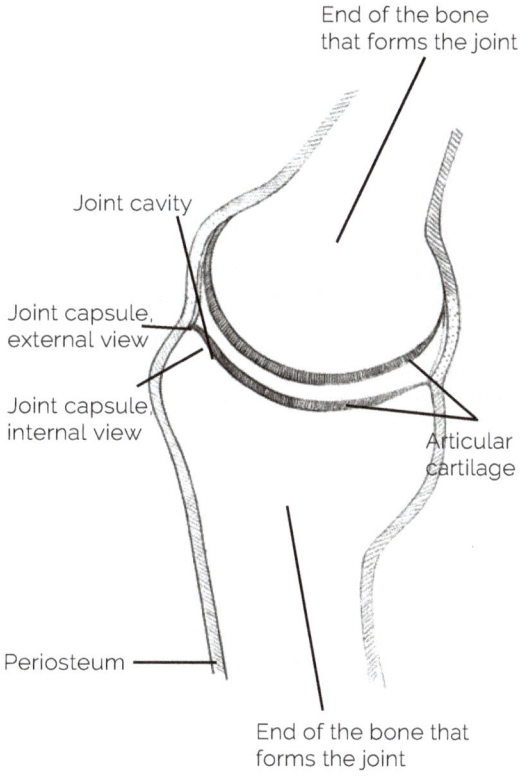

Suture between the halves of the frontal bone.

Did you know ...?

Hyaline articular cartilage is very sensitive. Too little movement leads to reduced supply of nutrients and the cartilage cells die off. Too much movement, especially in young, insufficiently warmed up or overweight horses, destroys cartilage cells because of the high compressive load. Foreign bodies in the joint (chips) can also destroy the cartilage, as well as poisoning, for example by selenium. Articular cartilage loses its elasticity and ability to regenerate as age increases. It becomes opaque and brittle and the cartilage cells eventually die off. In horses, all cases of this are called "arthrosis".

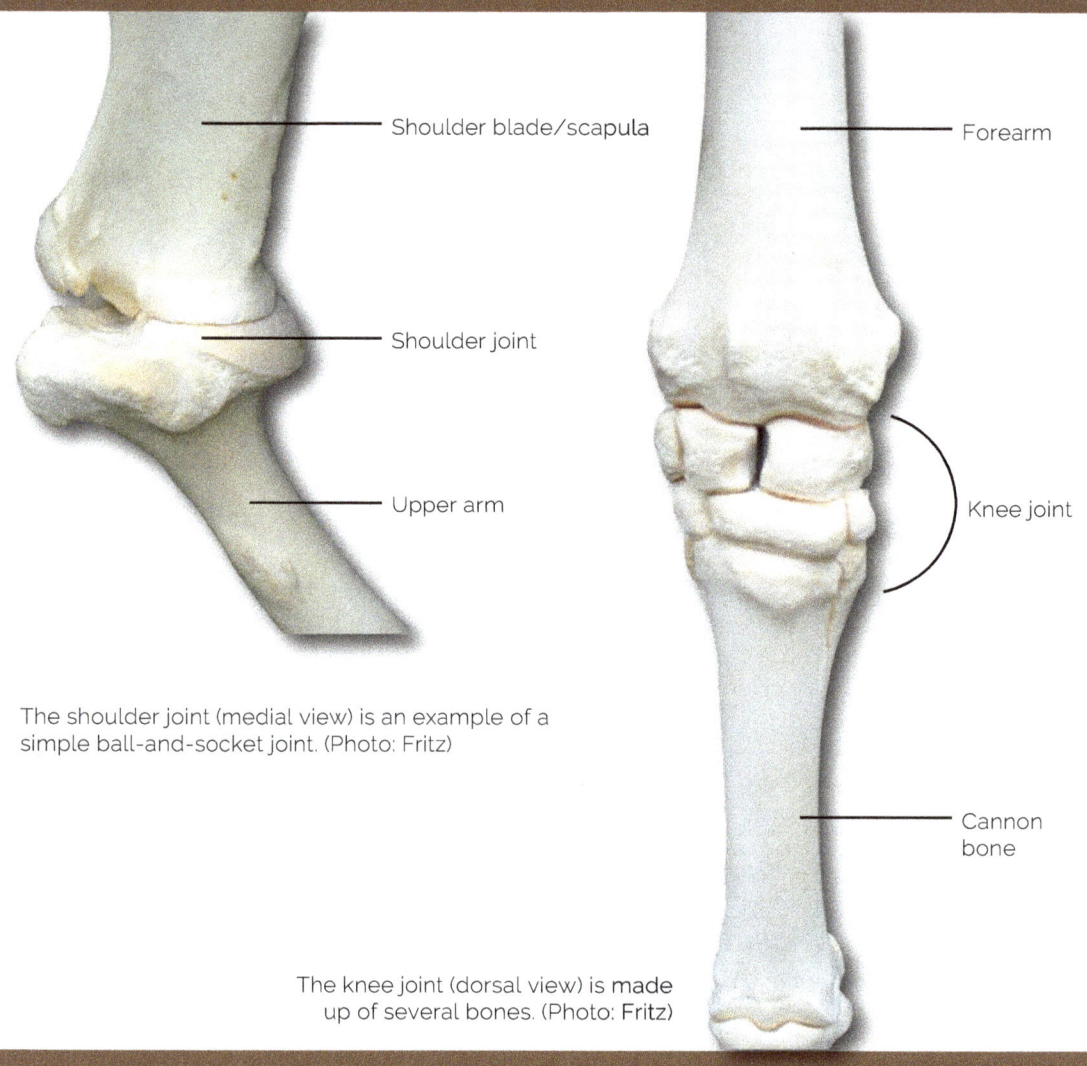

The shoulder joint (medial view) is an example of a simple ball-and-socket joint. (Photo: Fritz)

The knee joint (dorsal view) is made up of several bones. (Photo: Fritz)

The entire joint is encased by the fibrous joint capsule that arises from the periosteum and hermetically seals the joint cavity against the outside. The outer layer of the joint capsule is made of taut connective tissue and it is very firm and tough. It gives the joint stability and protects the joint cavity against injury, for example in the event of a fall. The inner layer of the joint capsule is traversed by blood vessels, nerves and lymph tracts. It is here that the synovial fluid originates, which is essential for lubricating and nourishing the joint capsule. There are articular ligaments inside and outside the joint capsule. They connect the ends of the bones together and stabilise the joint.

In some joints there are extra cartilaginous discs between the ends of the bones, which are covered with hyaline articular cartilage, for example the discus in the mandibular joint or the menisci in the stifle joint. While the In some joints there are extra cartilaginous discs between the ends of the bones, which are covered with hyaline articular cartilage, for example the discus in the mandibular joint or the menisci in the stifle joint. While the discus in the mandibular joint purely has a buffer effect, the wedge-shaped menisci help to stabilise the stifle joint.

There are different ways of categorising the joints of the musculoskeletal system. We can distinguish between the joints according to how many bones are involved in forming the joint. There are simple joints that are made up of just two bones, such as the shoulder joint for example, which, in horses, is formed by the shoulder blade and the upper arm. There are also compound joints that consist of several bones, such as the knee joint with its complex structure. However, joints are more commonly categorised by type of movement or function.

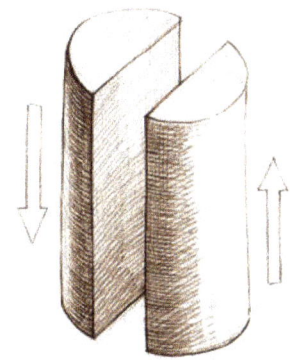

Sketch of a plane joint. (Illustration: Mähler)

The plane joint is one of the simplest joints. In this joint, the level contact surfaces of both bones involved glide against one another on one plane. This type of joint can be found in the vertebral joints, for example.

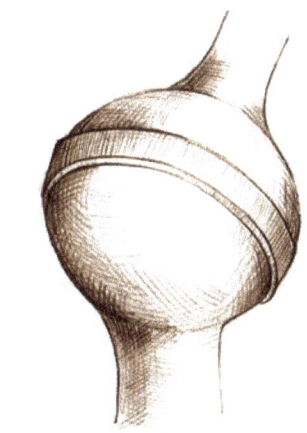

Sketch of a ball-and-socket joint. (Illustration: Mähler)

A ball-and-socket joint consists of two very differently shaped epiphyses. One of the bones is characterised by a spherical end, whereas its counterpart is shaped like a socket into which the ball fits exactly. The advantage of a ball-and-socket joint is the great freedom of movement in all directions, including rotational movements. However, this mobility may be restricted by muscles and ligaments, such as in a horse's hip joint, for example.

The condyloid joint looks similar to the ball-and-socket joint, but here the surfaces of the joint have an elliptical shape.Correspondingly, this joint allows a similar movement to the ball-and-socket joint. In horses, a condyloid joint can be found at the poll between the skull and the first cervical vertebrae.

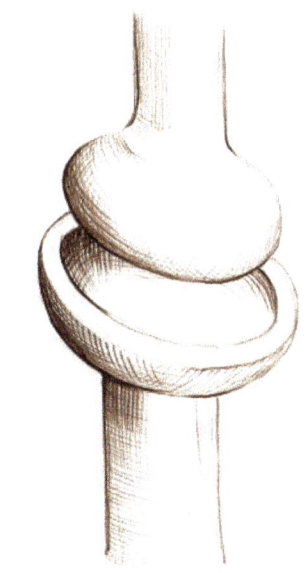

Sketch of a condyloid joint. (Illustration: Mähler)

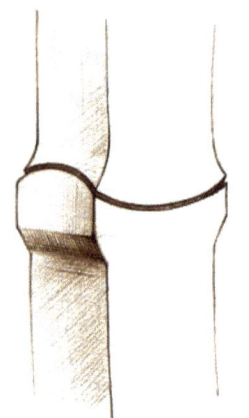

Sketch of a saddle joint. (Illustration: Mähler)

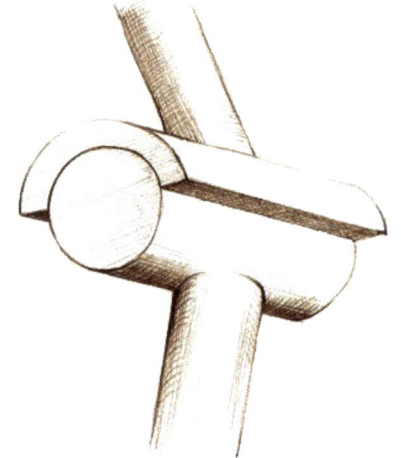

Sketch of a bicondylar joint (basic form). (Illustration: Mähler)

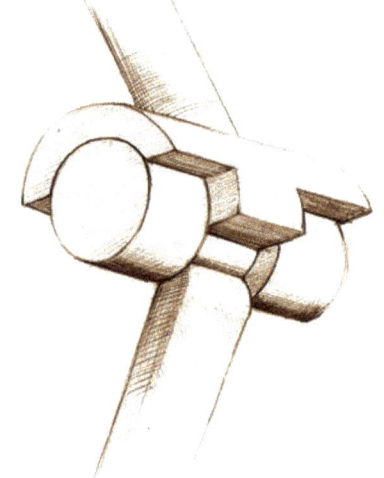

Sketch of a hinge joint. (Illustration: Mähler)

The saddle joint is most commonly found at the horse's extremities. In these joints, the joint ridge is shaped like the seat of a saddle. The socket fits the saddle shape. This joint mainly allows bending and stretching. Side-ways movement and rotation are usually also limited by ligaments.

The bicondylar joint looks a little like a transverse roller. Correspondingly, movement is only possible on the axis of the roller, i.e. a bending and stretching movement. Bicondylar joints are usually supported with strong ligaments to prevent them from twisting. Bicondylar joints can also be subdivided further into hinge joints, spiral joints, carriage joints and cochlear joints.

In a hinge joint, the roller-shaped joint ridge fits precisely into the relatively flat socket. This joint only allows bending and stretching. Hinge joints usually have additional pronounced guide combs that prevent any sideways movement or twisting. One example in horses is the elbow joint.

The spiral joint also has very pronounced guide combs, but here they are at an angle to the joint axis. The spiral joint therefore also only allows for bending and stretching, but both bones are turned in relation to one another during bending. A spiral joint can be found in the horse's hock (tarsocrural joint).

The carriage joint is also a bicondylar joint, even though it looks slightly different. In horses, the femoropatellar joint is a carriage joint. The patella is embedded in the patellar tendon of the large quadriceps femoris muscle and glides through a groove over the femur like a carriage on a rail.

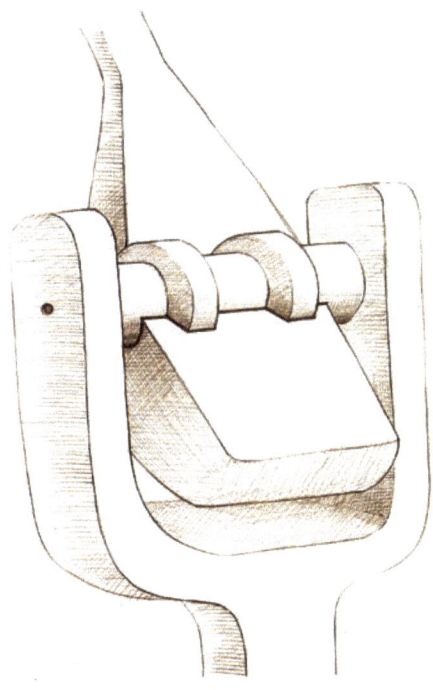

Sketch of a spiral joint. (Illustration: Mähler)

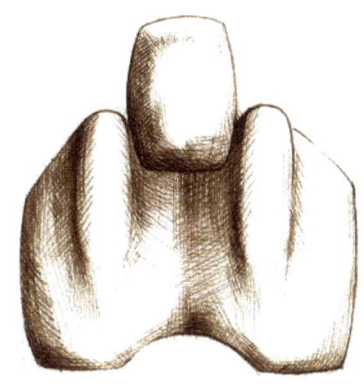

Sketch of a carriage joint. (Illustration: Mähler)

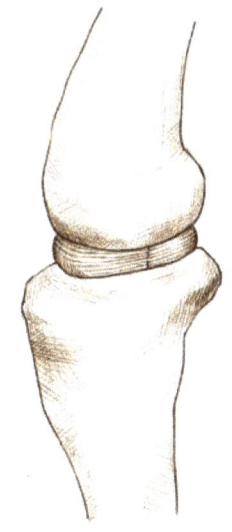

Sketch of a cochlear joint. (Illustration: Mähler)

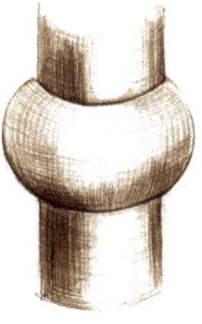

Sketch of a pivot joint. (Illustration: Mähler)

In terms of appearance, the cochlear joint resembles a spiral coil. Here the bones also move at an angle to one another, as in the spiral joint. The difference is that, in the cochlear joint, the ligaments are attached so that they inhibit the joint's movement by preventing it from being stretched beyond a certain angle. A cochlear joint of this kind can be found in the femorotibial joint where it prevents the joint from being damaged by excessive movement.

Like a bicondylar joint, a pivot joint only allows movement in one direction. However, movement is on a longitudinal axis and not a lateral axis. The joint pin of one bone projects into a hole in the opposite bone. This type of joint can be found, for example, between the first and the second cervical vertebrae of the horse, where it allows the head to pivot on the axis of the cerfical spine.

Another joint in the musculoskeletal system of a horse is the cartilaginous joint. In this

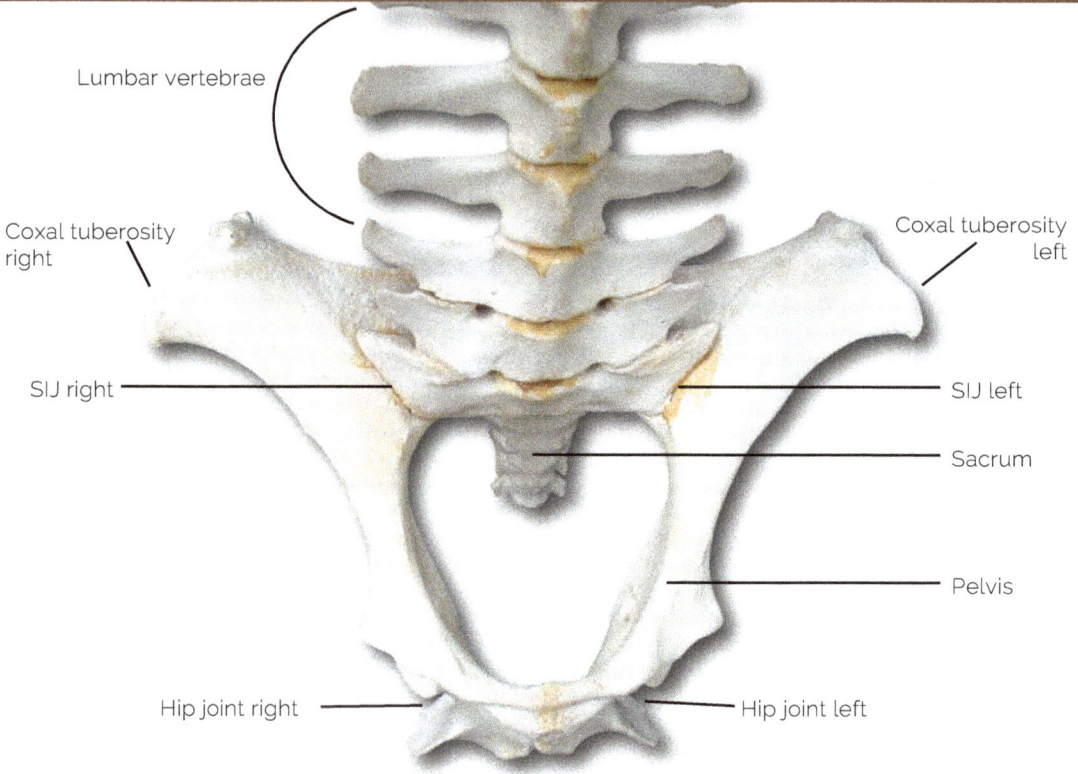

Pelvis and sacrum with sacroiliac joint, view from the abdominal cavity outwards. (Photo: Fritz)

joint, the relatively level ends of the bones lie flat on top of one another and are braced against each other with ligaments so that hardly any movement is possible. In horses, the sacroiliac joint, which attaches the pelvis to the sacrum, is cartilaginous joint. These joints are particularly sensitive to twisting movements because the short, tight ligaments allow them little room for manoeuvre.

The Skeleton

The bony skeleton of the horse is a marvel of construction made up of more than 200 individual bones that give the horse stability and create its shape. Each and every bone in the skeleton has a specific function and a shape and structure to match. Bones can move in relation to one another. This movement is made possible by the flexible parts of the musculoskeletal system, the joints, ligaments and musculature. The diagrams on page 23 give an overview of the parts of the skeleton that can easily be felt in horses.

The Spinal Column

Like us, horses are vertebrates. The spinal column is the central part of the skeleton. It provides stability but its structure, made from

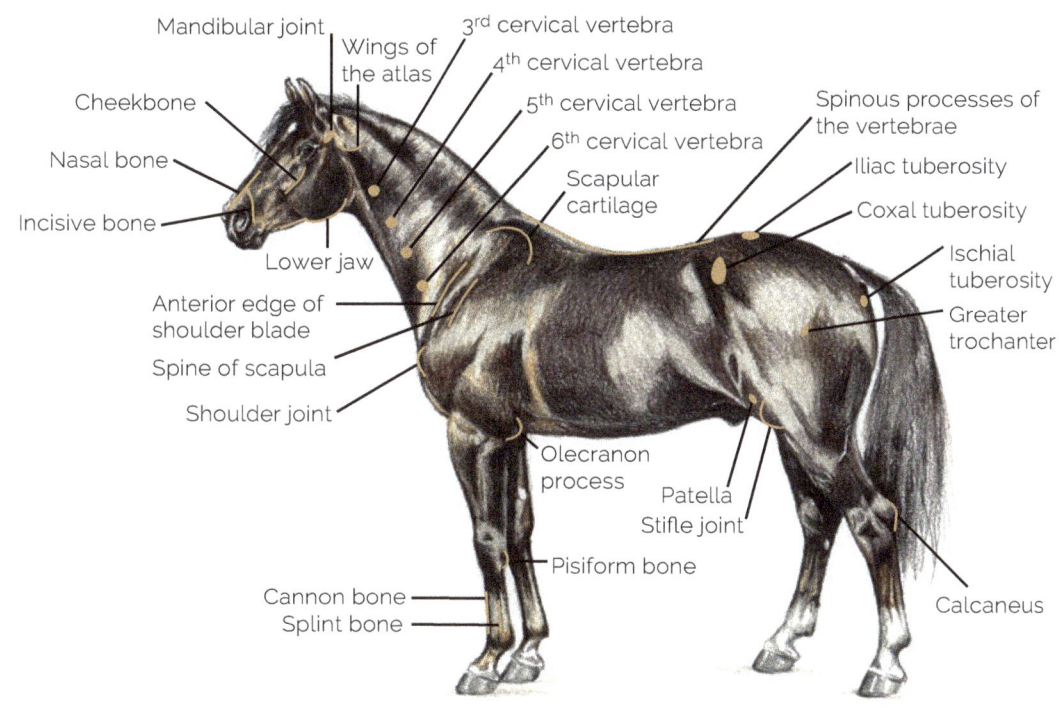

Points of the skeleton that can be felt on the surface. (Illustrations: Retsch-Amschler)

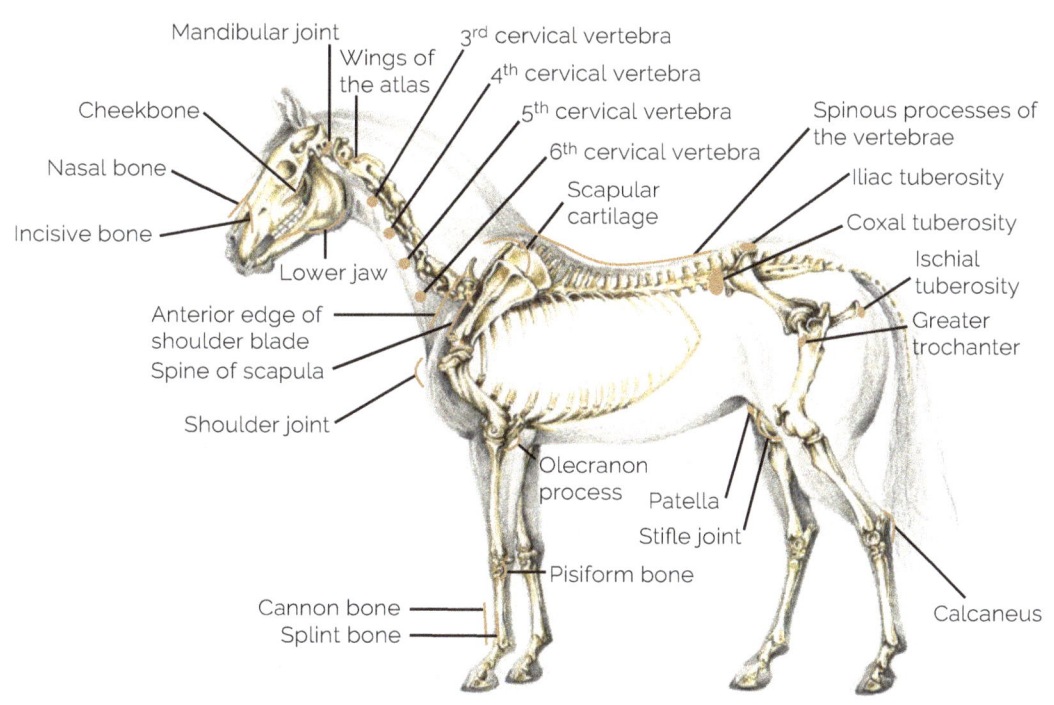

The Musculoskeletal System

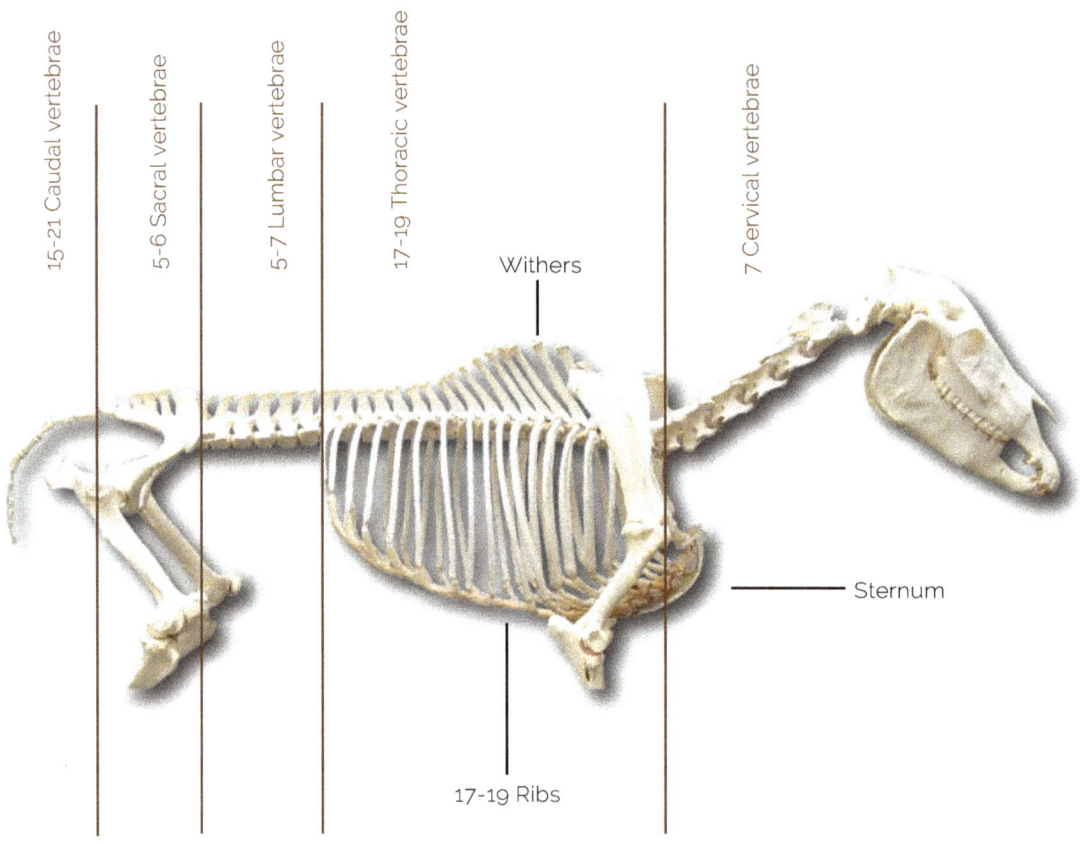

The spinal column of the horse can be divided up into five sections. (Photo: Fritz)

individual vertebrae, connected together so that they can articulate, means that it is also flexible, enabling the horse to bend to the left and right, as well as to hollow and hump its back. This flexibility is important for movement.

Like almost all mammals, the horse has seven cervical vertebrae that all have a slightly different shape. The first cervical vertebra is called atlas, after the mythological figure who bore the spheres of the heavens. It "carries" the head. The two wings of the atlas give it its shape. The second cerebral vertebra has a tooth-shaped process that reaches into the first cerebral vertebra (pivot joint). All of the movements of the head, such as turning, nodding or shaking originate here and at the condyloid joint between the head and the atlas, i.e. at the poll. The third to seventh cervical vertebrae all look very similar. On the whole, the cervical vertebrae are rather rounded and compact and do not have any long processes. This allows the neck its great freedom of movement. To touch, the vertebrae feel like large, firm rubber balls.

The cervical spinal column joins the thoracic spinal column at the thorax. In most horses, the thoracic spinal column consists of eighteen thoracic vertebrae, each of which supports a pair of ribs. Along with the ribs and the sternum, the thoracic spinal column forms the thorax that protects the heart and lungs. The thoracic vertebrae also all have spinous processes of different lengths that form the withers and that can be felt along the entire midline of the back. At the front, these spinous processes are aligned caudally, because the muscles attached here mainly pull in the direction of the head. The spinous processes point in the opposite direction so that they can easily resist this muscular strength. Shortly before the transition to the lumbar spinal column, the direction of the spinous processes is reversed and they then point in a cranial direction. The muscles that are attached to them pull caudally.

The thoracic spine becomes the lumbar spine roughly at the point where a general purpose saddle ends. The last rib can easily be felt by feeling the flank from the back to the front. If you follow this last rib up, you will come to the transition between the thoracic and lumbar spines. The bony basis for this last section of the back is six lumbar vertebrae, usually just five in Arabian horses, with very wide costal processes. While the thoracic spinal column is stabilised by the ribs and sternum, the lumbar spinal column is only held and secured by muscles.

The sacral vertebrae that join onto the lumbar vertebrae are fused onto the sacrum and form the osseous connection with the pelvis, the sacroiliac joint or SIJ for short. The sacrum only becomes completely ossified at four to five years of age. We should not expect young horses to take too much weight on their hindquarters because the sacrum is not yet able to take the strain. The caudal spinal column joins on behind the sacrum. The vertebrae of the caudal spinal column are relatively small and compact and have neither vertebral canal nor processes. The horse has 15 to 21 caudal vertebrae that form the dock. All vertebrae

The structure of a vertebra, using the example of a lumbar vertebra. (Illustration: Retsch-Amschler)

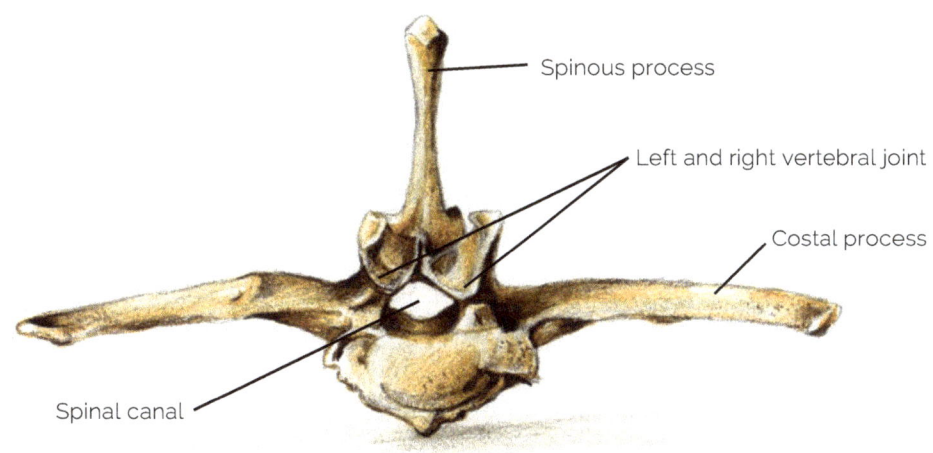

up to the sacrum have a recess at their centre for the spinal cord. Lined up next to one another, the recesses of the individual vertebrae form the vertebral canal in which the spinal cord runs. The vertebrae are joined together by two firm intervertebral joints each, which are also known as facet joints. These joints are located to the right and left sides of the vertebral canal and they are differently shaped in different areas of the spinal column.

The Skull

The horse's skull consists of 29 flexibly connected bones. However, there is only one true joint, the mandibular joint, which makes the chewing motion between the upper and lower jaw possible. All of the other connections are fine, fibrous seams called sutures. The sutures make it possible for the bones of the skull to move passively against one another. Restricting this movement, with an excessively tight browband or noseband for example, can cause headaches.

Via the occipital bone (os occipitale), the skull forms the poll with the first cervical vertebrae (atlas). The hard edge between the ears is called the nuchal crest and is part of the occipital bone. The nuchodorsal ligament is attached to it and extends along the entire spinal column right down to the sacrum. Together with the back muscles, the nuchodorsal ligament creates the counter tension to the abdominal muscles during ridden work.

The temporal bones (ossa temporale) are attached to the left and right of the occipital bone in a rostral direction. Along with part of the vault of the cranium and the right and left zygomatic arches, the temporal bones form the upper part of the mandibular joint. The os tympanicum is also embedded in the temporal bones. The vestibulocochlear organ (cochlea) and the vestibular apparatus (bony semicircular canals) are also found in this part of the bone (see also the section entitled "The Horse's Senses" page 90 onwards).

The bridge of the nose is formed by the nasal bone. The rostral part of the nasal bone is not attached and can easily be broken by incorrectly buckled headcollars or hackamores. The frontal sinuses are located below the nasal and frontal bone, roughly between the eyes. They are hollows in the skull through which respiratory air flows so that it can be warmed up, moistened and filtered free of dust. The ethmoid bone, which lies ventrally to the frontal sinuses, also performs this function. The frontal sinuses are connected left and right to the maxillary sinuses, which are located in the upper jaw bone. In young horses, the roots of the molars protrude into the maxillary sinuses.

The lacrimal bone (os lacrimale) is situated between the nasal bone and the upper jaw, at the height of the inside corner of the eye. The tear duct runs inside it. Tear fluid flows over the eye as a film, gathers in the inside corner and then drains through the tear duct into the nasal canal and then out. If the tear duct becomes blocked, the tear fluid drains out over the face, which we call "runny eyes".

The clearly palpable cheekbones located on the sides of the skull are formed by different bones. The upper jaw bone (os maxillare) forms the foremost part. The zygomatic bone (os zygomaticum) forms the part underneath the eye. The process behind the eye is formed by the zygomatic arch (arcus zygomaticus). The cheekbone is the point of contact for the large

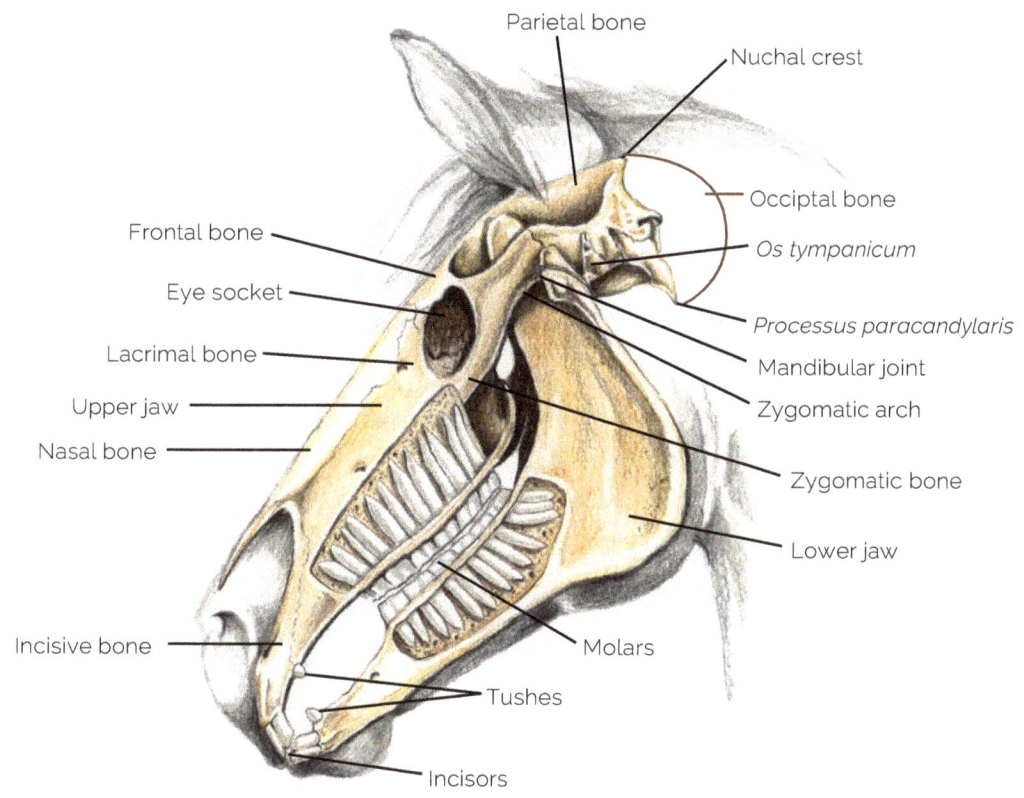

The horse's skull. (Illustration: Retsch-Amschler)

masticatory muscle (*M. masseter*). It is the strongest muscle in the body and can press the teeth together with a force of up to five tonnes.

The Teeth

The teeth are anchored into the upper jaw (*os maxillare*), the incisive bone (*os incisivum*) and the lower jaw (*os mandibulare*). Foals start off with milk teeth, some of which have already broken through the gums at birth. These milk teeth are then replaced by permanent teeth later on. The following rule of thumb applies to the incisors: the two middle incisors (centrals) can usually be seen at birth and change at around 2.5 years. The adjacent laterals come at around six weeks after birth and change at 3.5 years. The outermost incisors (corner teeth) appear at the age of around six months and then change at around 4.5 years. Once all of the milk teeth have been replaced by permanent teeth and when all of these teeth are fully pronounced, we can assume that the horse has finished growing. The permanent or true molars are not formed as milk teeth but appear as

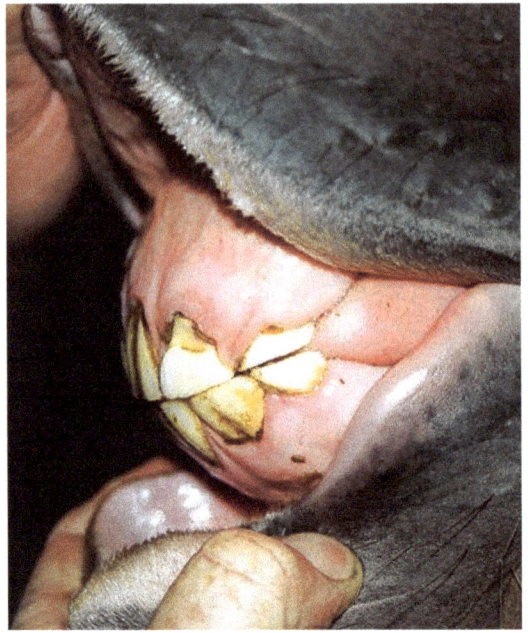

The corner teeth of this four-year-old are still milk teeth. The centrals and laterals have already changed. The remains of a milk tooth can still be seen on the upper lateral. (Photo: Kreling)

permanent teeth between one and three years of age so that foals have twelve molars and fully grown horses have 24.

Did you know ...?

Horses often suddenly develop mouth problems when they are being broken in, i.e. at between three and five years old, and resist the bit or the rider's hand. This is often caused by pain as a result of second dentition.

The horse has a tooth-free area between the incisors and the molars, which is known as the bars of the mouth (*diasthema*). This is where the bit lies during work, preferably directly in front of the molars. In stallions and geldings, the tushes are located at the front of the bars. In rare cases, mares can also have tushes. Wolf teeth may occur in the bars, regardless of gender. They are extra molars that have reduced in size during the course of evolution. Wolf teeth may be visible or only noticeable by touch. In the bit area, wolf teeth may cause pain. If this is the case they should be removed by an equine dentist.

Teeth are made from two different materials: soft dental cement and very hard dental enamel. Whereas in dogs, for example, the soft cement is on the inside and the hard enamel is on the outside, in horses enamel and cement are alternately folded into one another. In young horses, all of the teeth are very long and are embedded deep into the bone. The soft dental cement is quickly worn down by hard grass stems and hay into indentations (*infundibulum*), while the dental enamel remains and forms a sharp edge. This creates a grinding surface that the horse can use to grind up even the hardest fibres such as bark or roots. Dental enamel also wears down over time so teeth will become two to three millimetres shorter every year, which is why the teeth slide further and further out of the jaw over the course of the horse's lifetime, in order to allow normal chewing. Only in very old horses do the teeth lose their grip and, at some point, fall out of the jaw bone.

The degree of wear on the incisors can tell us a horse's approximate age. However,

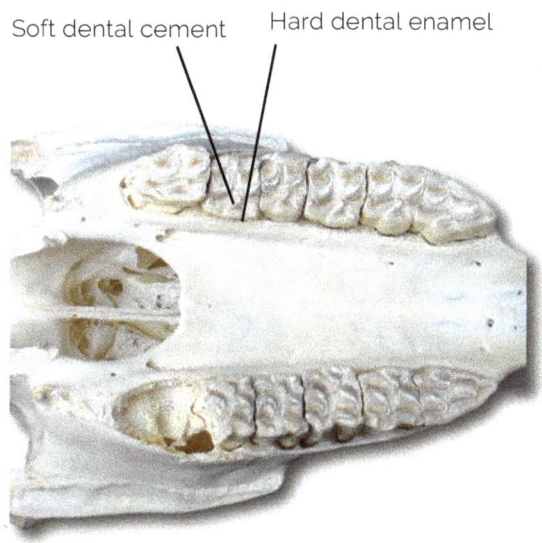

The upper molars of a horse, viewed from the oral cavity. (Photo: Fritz)

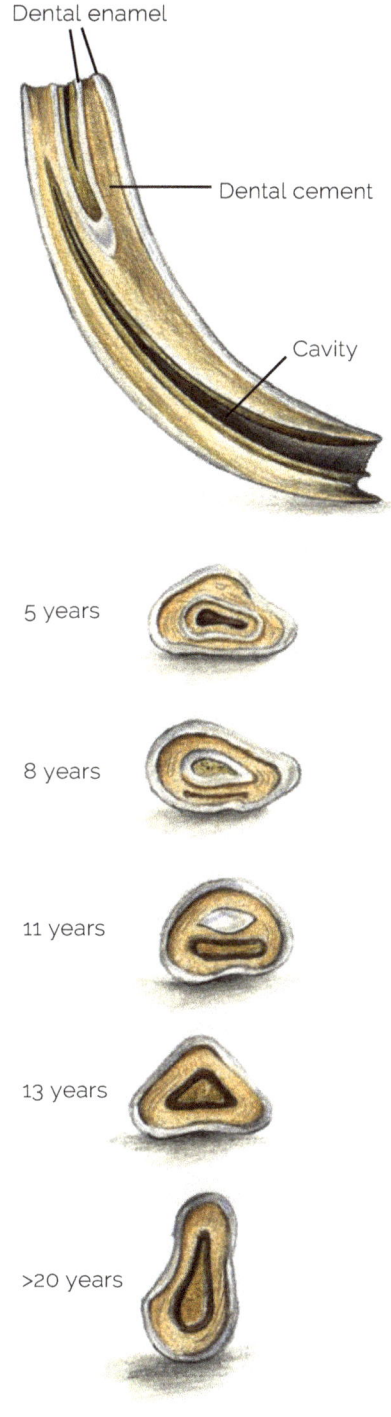

Cross-section of an incisor and changes to the tables caused by age-related wear. (Illustration: Retsch-Amschler)

the condition of the teeth depends on nutrition. The more a horse grazes and the harder and more copious the roughage he consumes, the faster the incisors will wear down, because the horse can still use them to bite off even the shortest stems of grass, as well as wood, bark and roots. If the horse is not able to wear down his teeth, an equine dentist will have to shorten the incisors regularly to allow for normal chewing. The molars grind down feed into the smallest fibres that should be smaller than 2 millimetres in a healthy mouth. Only thoroughly chewed food can be adequately digested, so well functioning teeth are the basis for optimum use of food and a sound metabolism.

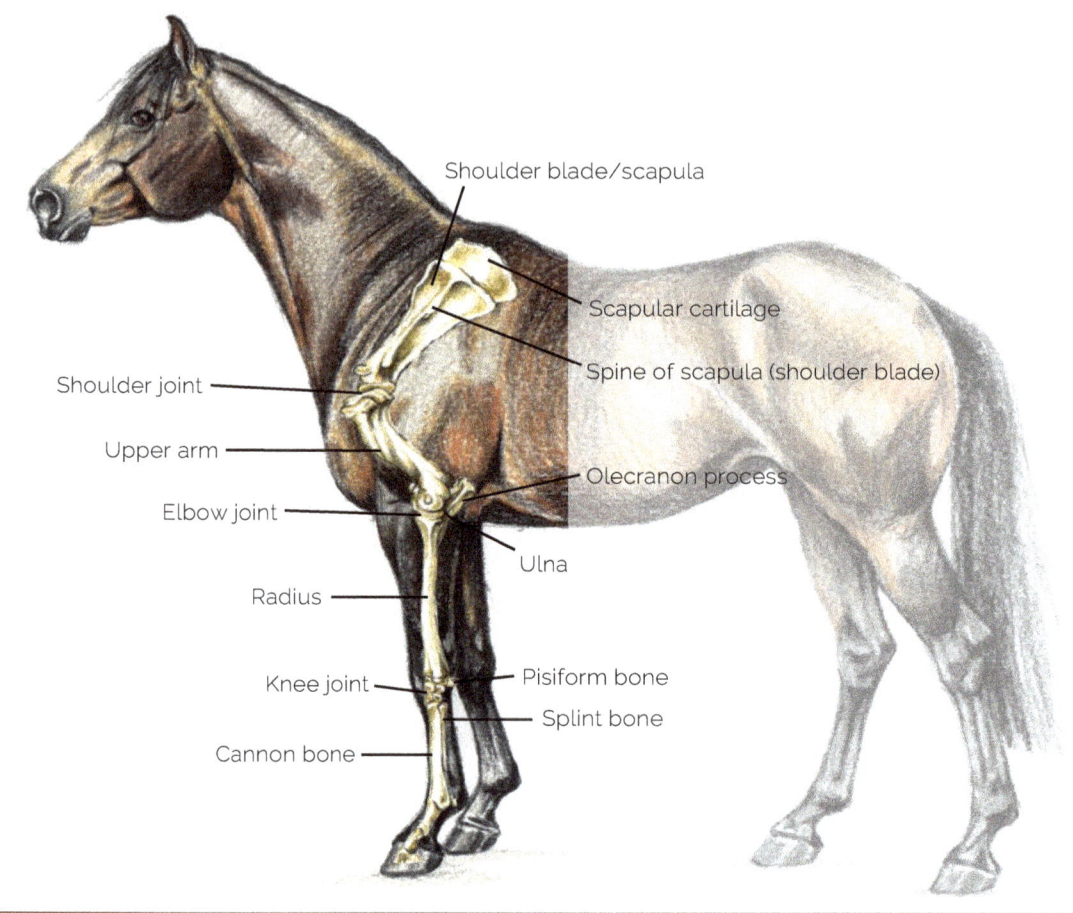

The horse's foreleg. (Diagram: Retsch-Amschler)

The Limbs

The bones of the limbs correspond to those of a human being if they were to walk on all fours. However, some changes have taken place during the course of evolution because of the horse's way of life, for example horses are unguligrades and not plantigrades like us. In order to achieve this, the bones of the fingers and toes have extended and then all reduced in size, apart from the middle finger or middle toe.

If you look at the foreleg, it begins in its proximal part with the shoulder blade (*os scapula*). Unlike in the human shoulder, the trunk is not joined to the foreleg with bone, so the horse also does not have a collar bone. Instead, the thorax is elastically suspended between the shoulder blades with muscles and fasciae. This construction gives the shoulder an important shock-absorbing function.

An overview of the bones and joints of the foreleg

Shoulder blade (scapula) *(os scapula)*
Flat bone with a clearly palpable cranial edge. The spine of the scapula is approximately central and is the point of influence for many muscles that allow the shoulder blade to tip forwards or backwards around its central fulcrum, depending on the movement. Dorsally pronounced cartilage cap which must not be crushed. The easily palpable shoulder joint is ventral.

Shoulder joint *(articulatio humeri)*
Ball-and-socket joint that is restricted by muscles and ligaments so that only bending and stretching are possible. The entire shoulder blade lifts away from the body during medial or lateral movements.

Upper arm *(os humerus)*
Short, thick tubular bone that is hidden almost completely beneath the musculature of the upper arm. Only the end points, i.e. at the shoulder joint and the elbow joint, can be felt.

Elbow joint *(articulatio cubiti)*
Hinge joint with snap effect, formed by the ulna, radius and the forearm. The olecranon process mechanically prevents the joint from over-stretching.

Forearm *(os radius)*
Unlike the human forearm, the equine forearm is formed by the radius *(os radius)* only. The ulna *(os ulna)* is stunted and has grown together with the radius in the distal area. It forms the olecranon process *(tuber olecrani)* that can easily be felt protruding caudally on the elbow joint. The underarm itself is underneath the flexor and extensor muscles.

Carpal joint/knee joint *(articulatio manus)*
Compound joint with three rows of joints, where movement is created by two bicondylar joints and a fixed row of joints. Palmar there is a palpable bony process, the pisiform bone *(os carpi accessorium or os pisiforme)*, which gives the flexor tendons a guide over the joint.

Below the knee, the foreleg has the same structure as the hind leg below the hock. We will go into the precise structure of the distal limb and the hoof in the section "Structure of the Distal Limb and Hoof".

Did you know …?

The movement of the shoulder blades must not be affected by the saddle. Western saddles are often fitted with the fork directly on top of the scapular cartilage with the result that the horse avoids movement from the shoulder. Many jumping saddles also restrict the movement of the shoulder blade because they are positioned too far forward or are too forward-cut.

If the foreleg is moved outwards, the whole shoulder blade lifts up. (Photo: Maurer)

The equine thorax is freely suspended between the shoulder blades. (Photo: Fritz)

32   A Journey through the Horse's Body

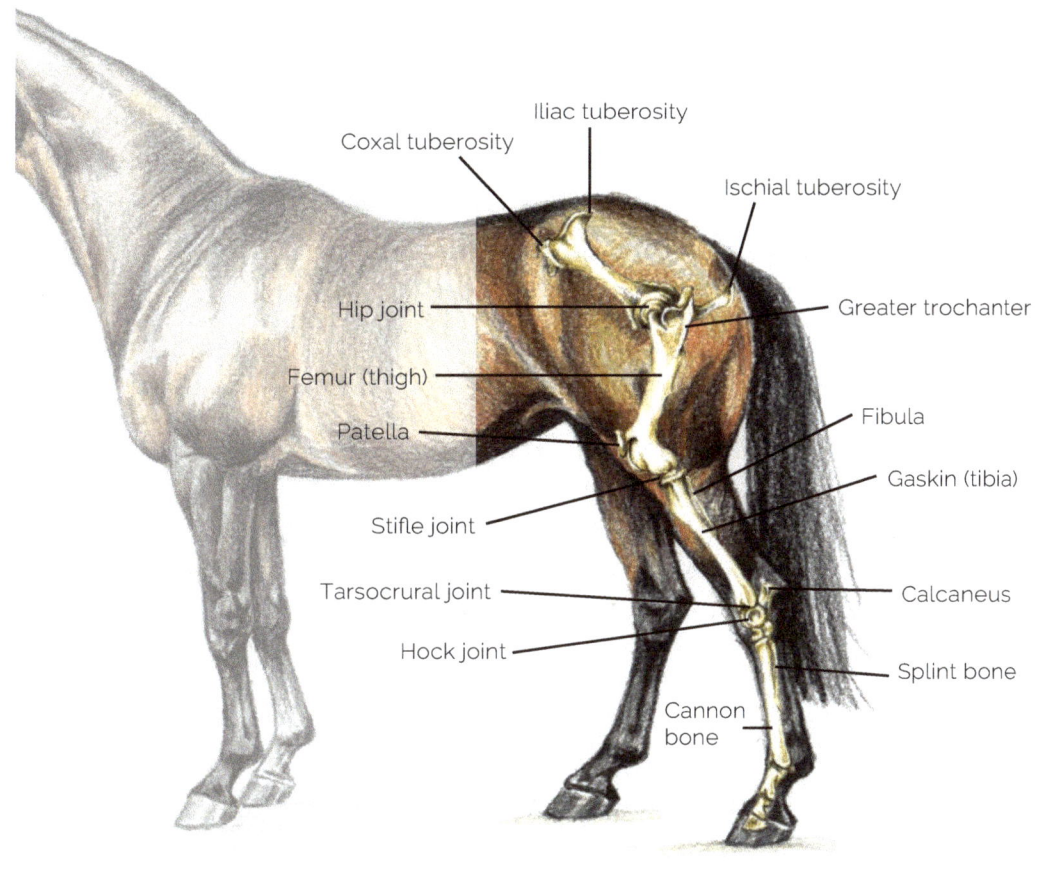

The horse's hind leg. (Illustration: Retsch-Amschler)

The hindquarters are the horse's engine that transfers its power to the forehand via the spinal column and the muscles of the back. Unlike the forelegs, the hind legs are joined to the skeleton of the trunk by bone. This connection is made by the pelvis, the sacroiliac joint and the sacrum. Here, the iliac wings of the pelvis lie flat on the sacral ala. The joint created acts like a shock absorber so that concussion from the back hooves hitting the ground is not transferred to the spinal column. These shocks are also absorbed by the angled knee and hip joints. The angles of the hindquarters play an important role in strength development in the hindquarters during riding. The more obtuse the angle; the less the horse is able to step under his centre of gravity.

The Musculoskeletal System

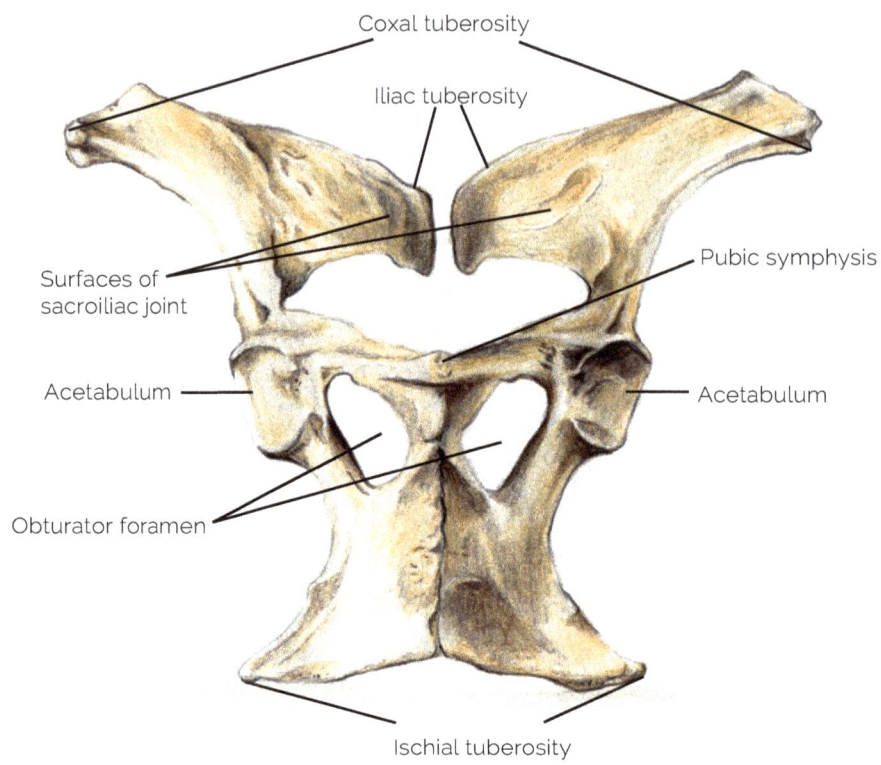

An overview of the bones and joints of the hind leg

Pelvis   (os pelvis)

The hind leg starts at the ring-shaped pelvis, which is composed of several flat bones. The large, flat ilium (*os ilium*) forms the joint to the sacrum (sacroiliac joint, SIJ). Dorsally, we can feel the iliac tuberosity (*tuber sacrale*), which forms the highest point of the croup and is particularly visible in thin horses. The easily palpable coxal tuberosity (*tuber coxae*) lies to the side. The ilium also forms part of the acetabulum. The acetabulum is also formed by the pubic bone (*os pubis*) that connects to the other half of the pelvis via the symphysis. Caudally, the acetabulum is formed by the ischium (*os ischii*), the hindmost part of the horse's skeleton. In most horses, the ischial tuberosities (*tuber ischiadicum*) can easily be felt underneath the tail, to the sides of the anus.

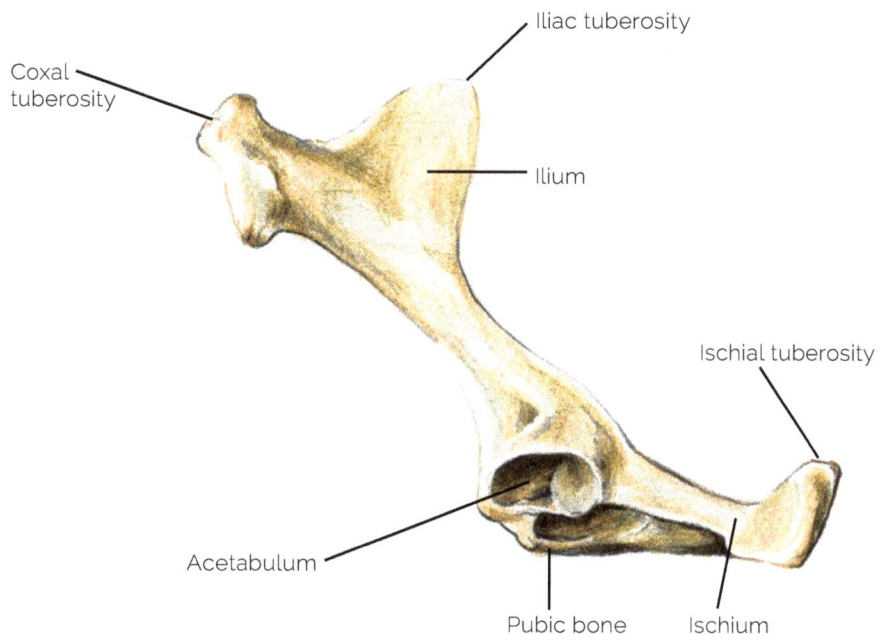

Pelvis, overhead view from ventral and view from lateral. (Illustrations: Retsch-Amschler)

Hip joint *(articulatio coxae)*
Ball-and-socket joint from acetabulum and the hemispheric articular head of the femur *(os femoris)*. In horses, it is held by very strong ligaments and muscles in a way that allows for bending and stretching but not circular movement. The hip joint itself is buried deep beneath the musculature. The greater trochanter of the femur can be felt at the hip joint.

Femur *(os femoris)*
Shorter, powerful long bone that is hidden under the musculature of the thigh. It ends distally in two hemispheric prominences of bone, called condyles, which form the stifle joint with the gaskin (tibia).

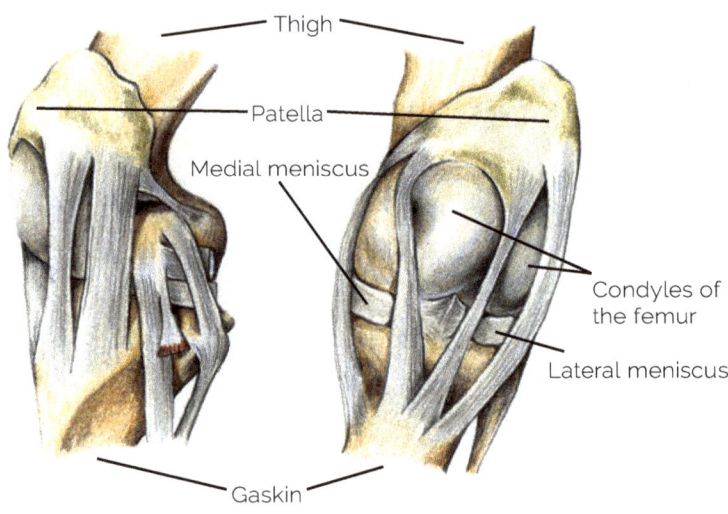

Stifle, lateral view and aerial view with stifle ligaments. (Illustration: Retsch-Amschler)

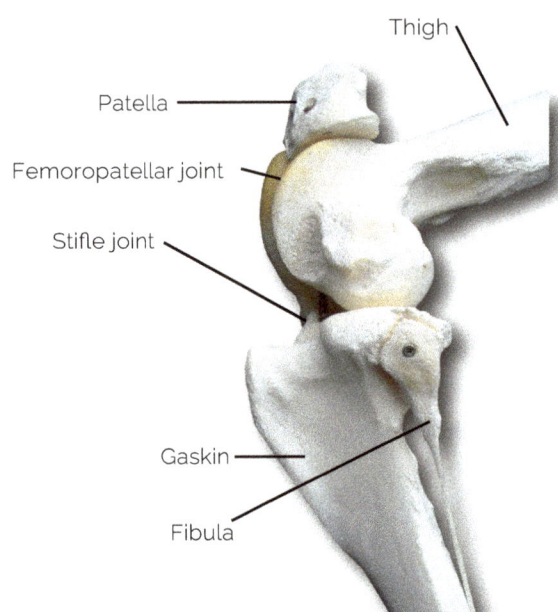

The bony parts of the stifle joint. (Photo: Fritz)

An overview of the bones and joints of the hind leg

Stifle (femorotibial joint: articulatio femorotibialis and femoropatellar joint: articulatio femoropatellaris)

The stifle is actually made up of two joints. The femorotibial joint is actually an incongruent joint, which means that two bones come together that do not actually fit. They are made to fit by the incorporation of two cartilage discs (*menisci*) and various retinacula. The stability of the stifle is mainly achieved by the muscular tension of the hindquarter and a correspondingly acute stifle angle. The femoropatellar joint is a carriage joint where the knee cap (*patella*) slides over the femur in order to redirect the patellar tendon over the knee.

Gaskin (*os tibia*)

Like the forearm, the gaskin or second thigh was originally two bones, namely the tibia (*os tibia*) and the fibula (*os fibula*). In horses, the fibula is stunted and has partly fused with the tibia. The bone is concealed beneath the flexor and extensor muscles.

Hock joint (*articulatio tarsocruralis*)

Like the knee joint, it is made up of several bones. The bend only takes place at the tarsocrural joint that is formed by the gaskin and the topmost bone of the hock joint (*os talus*). There are a further two rows of bones distally on the os talus with firm joints that hardly exhibit any mobility and that frequently become inflamed and then ossify under the strain of high impact. This process is described as bone spavin. The calcaneus (*os calcaneus*) projects caudally out of the hock. It is a deviation point for the flexor tendons that run down the plantar side of the hind leg. Mechanical irritation may cause thickenings to occur at this point (bursitis, also known as capped hock).

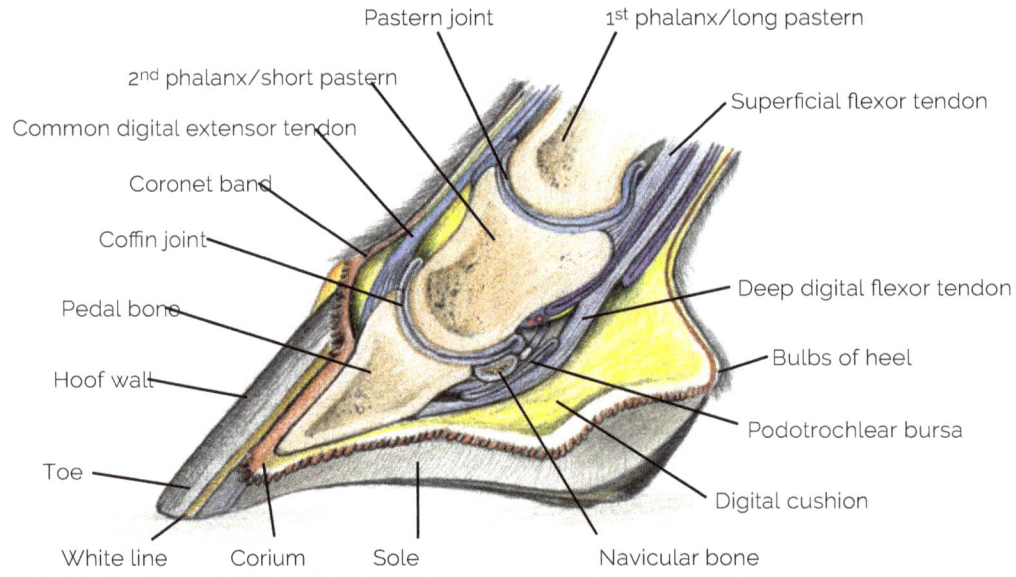

The hoof structure in detail. (Illustration: Retsch-Amschler)

Structure of the Distal Limb and Hoof

Distally from the knee or hock joint, the horse's legs have an almost identical structure. The bones and joints are the same, but the shape and angles may differ very slightly. Furthermore, all of the muscles end above the knee or hock joint, so movement and stability in the distal section of the leg are only created by tendons and ligaments. This has a clear effect on circulation in the legs because blood flow in muscular tissue is considerably greater than in bone, ligament or tendon tissue. That is why muscular injuries usually heal much faster than tendon or suspensory ligament injuries.

Let's look a little more closely at the anatomy of the hoof. The pedal bone is encased by the hoof capsule that forms the hoof wall on the outside and the sole underneath. A "white line" is created where the corium and the hoof wall meet. It can be seen easily when the hoof is freshly pared. At this point, the hoof capsule and the dermis are joined together with extremely fine lamellae. The podotrochlear apparatus is located on the sole side between the digital cushion and the phalanges. It consists of the navicular bone, the podotrochlear bursa and the deep digital flexor tendon. The digital cushion has a shock-absorbing effect and acts as a pump mechanism for supplying the hoof. The coronet band is where the skin and the hoof horn meet. Specialised cells in the coronet band produce the horn that forms the hoof wall. Damage to these cells, caused by a kick to the coronet band, for example, leads to the formation of weak hoof horn at this point and therefore makes the horn capsule easier to tear.

An overview of the bones and joints of the distal limbs

Cannon bone *(os metacarpale III)*
Greatly extended and thickened middle finger/toe. The cannon bone has to carry a great deal of weight and therefore has a particularly stable construction. The splint bones (*ossa metacarpale II and IV*), lie at the side of the cannon bone. They are the stunted index/ring finger bones and the corresponding toe bones. The splint bones act as guide rails for the flexor tendons on the back of the leg.

Fetlock *(articulatio metacarpophalangea)*
Hinge joint with a pronounced ridge that prevents rotation. The fetlock joint has a shock-absorbing function. When the horse takes a step, the fetlock joint sinks towards the ground and the impact energy is absorbed by the flexor tendons. The sesamoid bones are located on the palmar/plantar side. They act as guide rollers for the flexor tendons and are held in place by a total of seven ligaments.

Long pastern/ 1st phalanx *(phalanx proximalis)*
Short bone that forms the part of the horse that we generally know as the "pastern".

Pastern joint *(articulatio interphalangea proximalis)*
Saddle joint located just above the coronet band, which, in addition to bending and stretching, also allows a slight rocking function, in order to compensate for uneven ground.

Short pastern/ 2nd phalanx *(phalanx media)*
Short bone that projects into the hoof capsule.

Coffin joint *(articulatio interphalangea distalis)*
Saddle joint to the pedal bone that allows bending, stretching and slight rocking movements. Located within the hoof capsule, the navicular bone (*os naviculare*) lies palmar/plantar at the coffin joint and forms the guide roller for the deep digital flexor tendon.

Pedal bone/ 3rd phalanx *(phalanx distalis)*
Hoof-shaped bone that is suspended within the hoof capsule above the corium. This suspension creates effective shock absorption as far down as the hoof area.

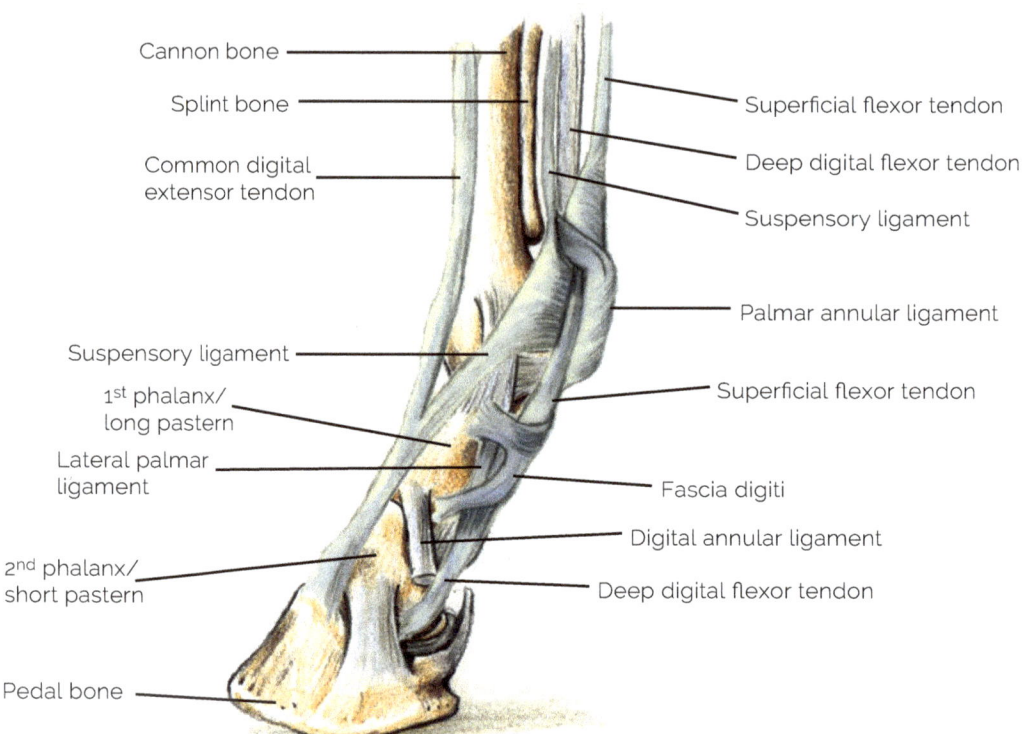

Distal limb with bones and ligaments. (Illustration: Retsch-Amschler)

The tendon and ligament apparatus of the equine distal limbs is particularly complex. The extensor tendons that extend to the hoof, or rather to the long pastern run on the dorsal side of the leg. By contraction of the extensor muscles that belong to it, the leg is stretched down to the toe. With regard to the flexor tendons, we differentiate between superficial and deep, depending on how they run along the palmar or plantar side of the cannon bone. If we elevate the leg so that there is no tension on the tendons, we can feel a thick cord, in which two tendons can clearly be felt running parallel to each other. The flexor tendons intersect at the pastern, so that the superficial flexor tendon ends in the distal area of the long pastern and at the proximal edge of the short pastern. The deep digital flexor tendon, on the other hand, continues down and ends on the bottom of the pedal bone. Therefore, the superficial flexor tendon flexes the fetlock and pastern joints and the deep digital flexor tendon also flexes the coffin joint.

A further, tendinous structure, the suspensory ligament (*m. interosseus medius*), can be felt between the flexor tendons and the cannon bone. It splits up just before the fetlock joint, goes around the back of the fetlock and its branches run in a dorsal direction where they merge with the common extensor tendon. So that the flexor tendons do not slip out of position when they move, they are held by the very powerful palmar annular ligament, which is found at the fetlock. Furthermore, they are also held in the pastern region by the fascia digiti, which attaches to the top and bottom of the long pastern. The digital annular ligament, a ligament that makes a ring around the flexor tendon from the long pastern, is located shortly before the point at which the deep digital flexor tendon enters the hoof capsule. The tendons are protected by tendon sheaths so that they are not torn and damaged by the various retinacula. They lie around the tendons like tubes and cushion them against irritation. These tendon sheaths may become inflamed and puffy under excessive strain. If this happens they usually ooze out of the retinicula above the fetlock, where they are called "windgalls".

The horse's hoof is a biomechanical marvel. It is simultaneously stable and flexible. Not only can it withstand concussion, it can also absorb it. It is light and constantly regenerates itself. At the same time, the hoof horn adapts to the conditions underfoot and harder horn is produced for harder ground. Furthermore, the hoof is a counterpart to the heart. Blood that is pumped by the heart in the direction of the leg is pumped back towards the heart as venous blood by the frog. This happens because the hoof expands – the front hooves sideways and the back hooves to the front and back - every time the horse puts its foot down and then contracts again when the foot is lifted. When the horse puts its foot down, the normally convex sole flattens out and the frog touches the ground, which again stimulates circulation and horn growth. This pumping action is crucial for good circulation and lymph drainage in the legs. Serious illnesses, metabolic disturbances and changes in feed are all reflected in the formation of hoof horn and can cause rings to appear on the hooves.

Hoof with rings. (Photo: Fritz)

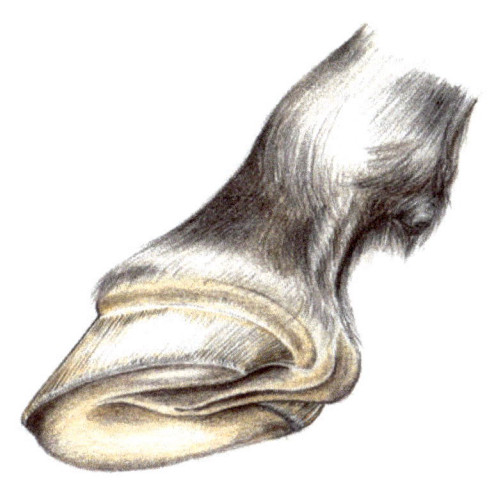

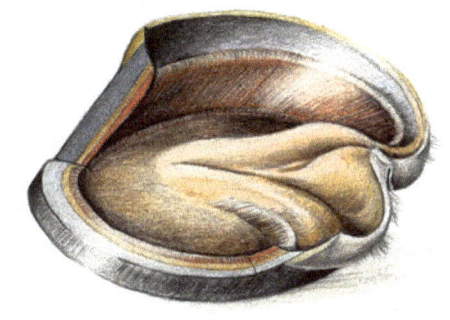

Unshod hoof and the hoof capsule belonging to it. (Illustrations: Retsch-Amschler)

The Musculoskeletal System 41

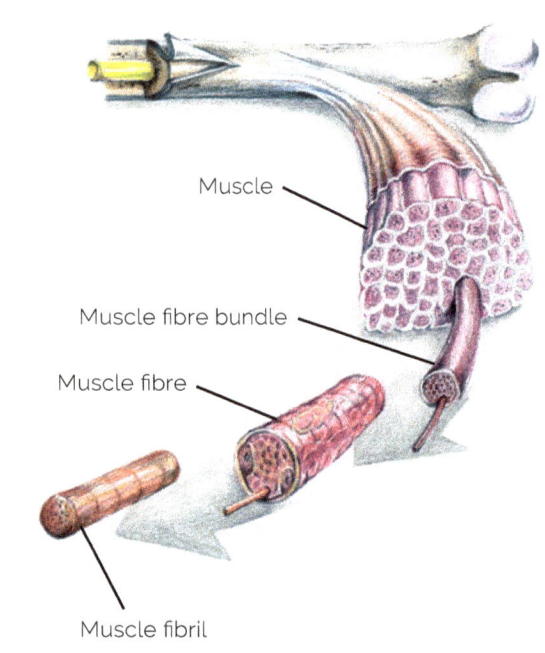

Each muscle is made up of many bundles of fibres that, in turn, are composed of muscle fibres. (Illustration: Retsch-Amschler)

The Musculature

The musculature makes up a large part of the body mass. Muscles are used for movement, on the one hand for locomotion and movement to adopt various postures and, on the other, for movement that supports the organs. In addition, muscles also stabilise and form the walls of the body.

A muscle consists of many bundles of muscle fibre that, in turn, are made of muscle fibres. Individual muscle fibres are made of thin muscle fibrils. These muscle fibrils contain sarcomeres, which are small, contractible units. During contraction, the thread-like filaments in the sarcomeres move past each other in opposite directions, rather like a group of people pulling on a rope: grasp, pull, grasp again, pull, shortening the rope all the time. Unlike pulling a rope, the filaments also release energy.

Muscle Types and Muscle Function

If we look at muscles under a microscope, we can essentially differentiate between three types. Most muscles are skeletal muscles. They appear transversely striated under the microscope. These striations correspond to the sarcomeres, which are arranged in parallel.

The transversely striated skeletal muscles are normally voluntarily controlled. They are almost all arranged in pairs so, in horses, the same muscles are on the left and right. Each skeletal muscle always has an origin and a point of influence and these may be far apart. In most cases, the proximal point is the origin and the distal point is the point of influence. The muscle is normally not directly attached to the bone. Instead, muscles are joined to the bone at the point of origin by the short tendon of origin. This tendon consists of short tendinous fibres that come from the muscle and radiate into the bone as insertions. At the other end of the belly of the muscle, the muscle ends in a terminal tendon, which can be of various lengths, that is inserted into another bone at the point of influence. The

Smooth musculature

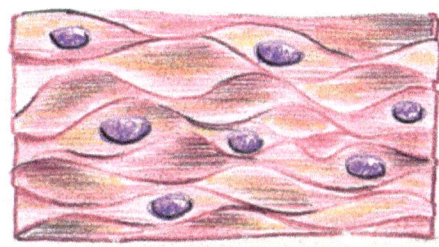

Transversely striated skeletal musculature

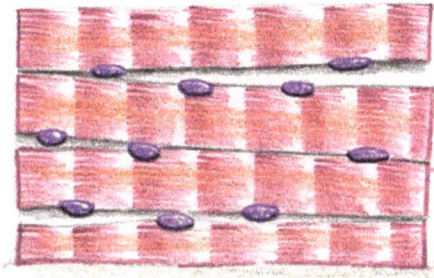

Transversely striated cardiac musculature

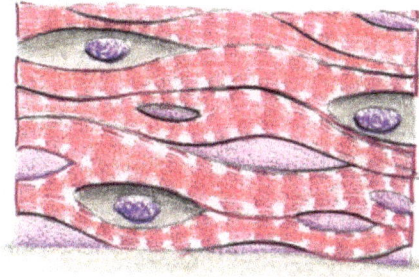

An overview of the different types of muscle (Illustration: Retsch-Amschler).

muscle between these tendons can contract and the tendons tighten and pull on it. The tendons themselves do not contract. The muscle either pulls directly on the bone via its origin or it transfers the pull to its terminal tendon and from there to bones further away. The pull causes the bones to move together. Tendons are encased in tendon sheaths to enable them to glide better, especially when they run over bony prominences. Where strain is particularly great, bones are embedded as guide rollers (sesamoid bones), for example the knee cap. There is fluid between the tendon and the tendon sheath that is very similar to synovial fluid. This type of fluid can also be found in the synovial bursae, which are like small, water-filled balloons that are used anywhere in the body where extra protection is needed. Synovial bursae are mainly found in joints where structures have to be protected against friction or pressure.

Some muscles have several muscle bellies and correspondingly several terminal tendons and points of influence. Some muscles also have several points of origin, for example, the tridentate "triceps". There are muscles that are interspersed with tendons. They can produce slightly less movement, but are more stable and can apply more strength as a result, for example the masticatory muscle. Some muscles originate from fasciae and not from bones. Fasciae are firm connective tissue membranes that stretch to form connective structures in the body. If these connective tissue membranes cover muscles to allow them to glide better against one another, they are called muscle fascia.

Most skeletal muscles have an antagonist that always performs the opposite movement. For example, a flexor

The cardiac musculature also has transverse striations, but is constructed in a slightly different way from skeletal muscle fibres. These muscle fibres cannot be influenced voluntarily and are constantly working throughout the organism's entire lifespan. Unlike the skeletal muscles, each individual muscle cell is not supplied by a nerve fibre. There are "gap junctions" between the cardiac muscle cells, across which electrical impulses can be carried. This will be covered in more detail in the chapter on the cardiovascular system.

There are muscle fibres without transverse striations, called smooth musculature, on the internal organs and the blood vessels. These smooth muscles take care of peristalsis and blood pressure regulation. This musculature can also not be influenced voluntarily.

Superficial, Medial and Deep Musculature of the Trunk

The various muscles that create a wide range of movements are arranged in layers. The superficial muscular layer is the one that we can feel by touch and directly influence with massage. The deep musculature of the trunk is mainly involved in stabilising the body and constantly sends information to the brain about the position of the body in space ("proprioception"). To give a better overview, the muscles and their many and varied purposes have been presented in the following table, where it becomes clear that most muscles have different functions, depending on whether they are fixed at the point of influence or at the origin and whether their partner on the other side of the body makes the same contraction or not.

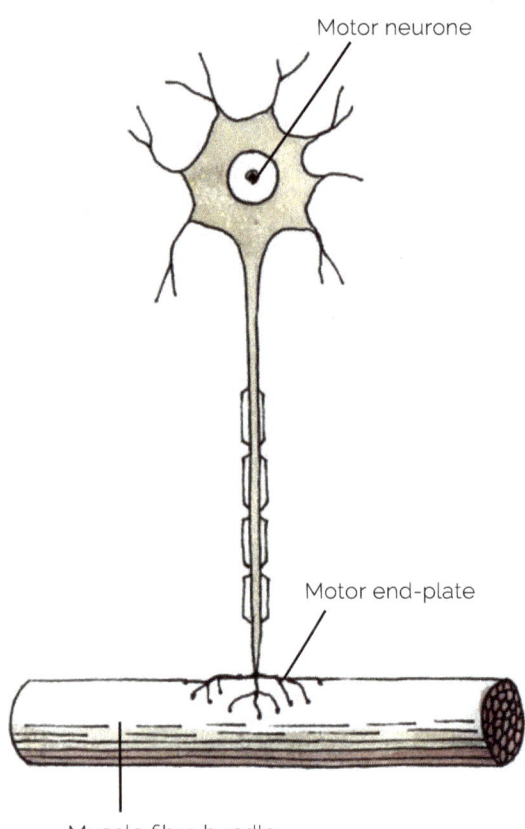

The motor end-plate makes the connection between nerve cell and muscle.
(Illustration: Mähler)

muscle in the leg is stretched when its antagonist, an extensor muscle, is tensed. Information about whether a muscle should be tensed comes from the nerves, which are connected to the muscle via the motor end-plate. Similarly, feedback about the current tonicity is sent to the brain via the nerves. In this way, the tension of each individual muscle in the body can be continually adjusted.

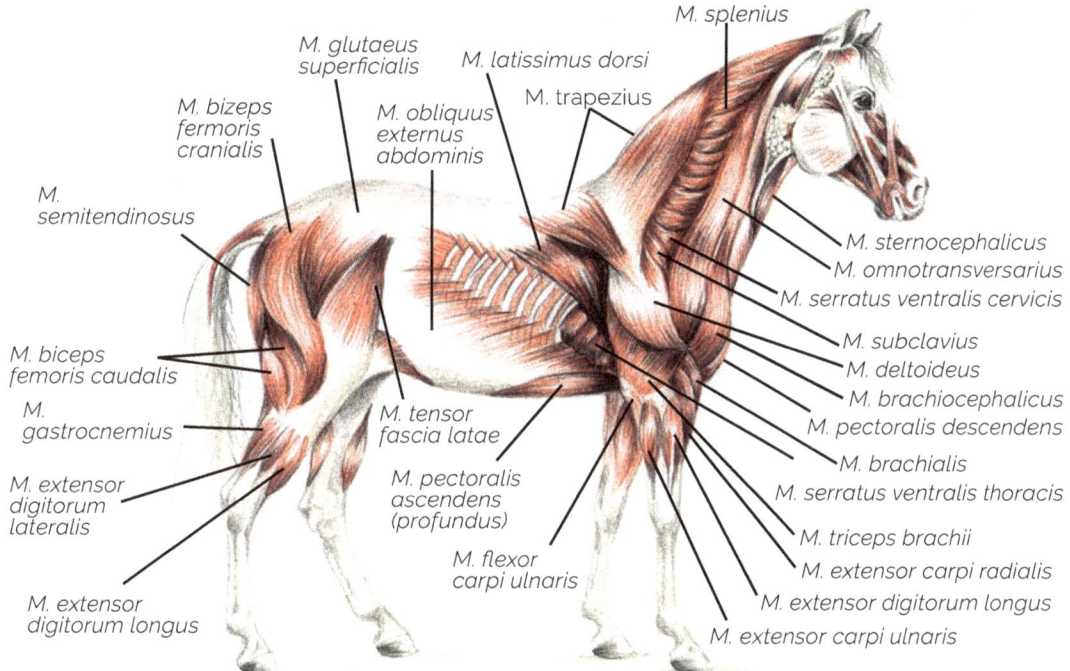

The superficial muscular layer. (Illustration: Retsch-Amschler)

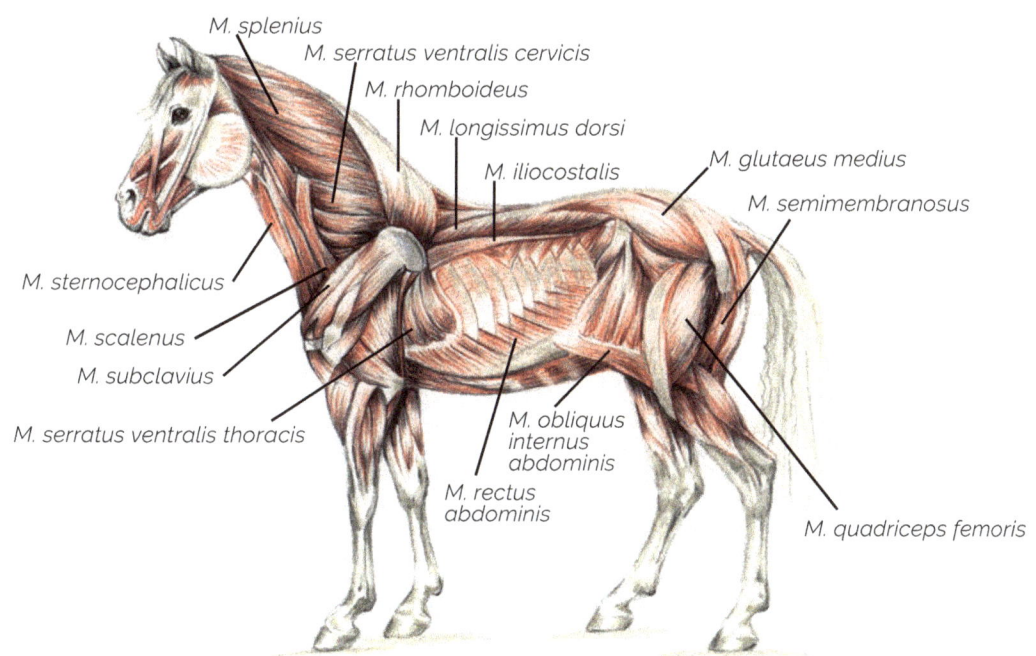

The medial muscular layer. (Illustration: Retsch-Amschler)

The Musculoskeletal System

An overview of the essential muscles of the neck and trunk and their functions

Muscles of the neck and poll

M. longissimus capitis	Stretch poll and neck if contracted bilaterally. Rotate and bend poll and neck laterally if contracted unilaterally.
M. sternocephalicus	Flexes poll and neck if contracted bilaterally. Rotates and flexes poll and neck laterally if contracted unilaterally.
M. brachiocephalicus	Fixed at the head: Forward movement of the forelimbs, determines gait and degree of collection, initiator of escape movement. Fixed at the upper arm: Extends the poll and flexes the neck when contracted bilaterally. Rotates and flexes head and poll laterally if contracted unilaterally.
M. scalenus	Fixed at the neck: Supports inspiration. Fixed at the rib: Flexes the base of the neck if contracted bilaterally. Rotates and flexes the base of the neck laterally if contracted unilaterally.
M. trapezius	Fixed at the shoulder blade: Involved in extension and lateral movement of the neck. Fixed at the nuchal ligament: Cranial and caudal movement of the proximal part of the shoulder blade during movement.
M. omotransversarius	Fixed at the shoulder blade: Supports extension of the neck if contracted bilaterally. Flexes the neck laterally if contracted unilaterally. Fixed at the neck: Moves the forelimbs forward, together with *M. brachiocephalicus*.
M. rhomboideus	Fixed at the neck: Movement of the proximal part of the shoulder blade in the direction of the head during the supporting phase. Fixed at the shoulder blade: Supports extension (bilateral contraction) or lateral flexion (unilateral contraction) of the neck.
M. serratus ventralis cervicis	Fixed at the neck: Movement of the proximal part of the shoulder blade in the direction of the head during the supporting phase. Fixed at the shoulder blade: Extends the base of the neck if contracted bilaterally. Rotates the base of the neck if contracted unilaterally.
M. splenius	Fixed at the cranial part: Extension of the trunk. Fixed at the caudal part: Extends the neck and raises the head and neck if contracted bilaterally. Rotation and lateral flexion of the neck if contracted unilaterally.
M. spinalis cervicis	Extends the cervical and thoracic spinal column.

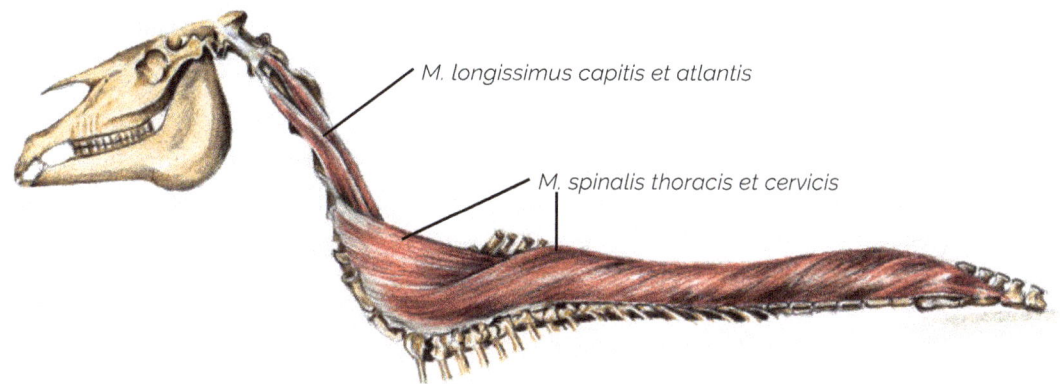

Erector of the spine *(M. erector spinae)*. (Illustration: Retsch-Amschler)

Muscles of the trunk

M. obliquus externus abdominis	Flexion of the thoracic and lumbar spinal columns when contracted bilaterally. Rotation and lateral flexion of the trunk if contracted unilaterally.
M. obliquus internus abdominis	Works synergistically with *M. obliquus externus abdominis* to flex the spinal column but works antagonistically during rotation.
M. rectus abdominis	Flexion of the thoracic and lumbar spinal columns when contracted bilaterally. Support of lateral movement and rotation when contracted unilaterally.
M. transversus abdominis	Maintains abdominal tension, works with all of the abdominal muscles mentioned above as an antagonist to the muscles of the back.
M. iliopsoas	Flexion of the lumbar region and the transition to the sacrum ("sitting the horse down") if contracted bilaterally. Rotation and slight lateral flexion of the lumbar region, bending of the hip joint and supination of the femur when contracted unilaterally.
M. iliocostalis	Important for breathing and lateral movement of the thoracic part.
M. spinalis	Extensor of the thoracic and lumbar spinal columns.
M. longissimus	Extension and lateral movement of the spinal column. Described as *M. erector spinae*, "erector of the spine" together with *M. spinalis* and *M. iliocostalis*.
Mm. multifidii	Attachment of the vertebral joints and the vertebrae to one another, proprioception (perception of the posture of the body). Involved in lateral movement and rotation of the spinal column.

The half-pass requires interplay between numerous different muscles. (Photo: Slawik)

In order to complete an entire cycle of movement, the leg section does not just have to be moved below the elbow by the flexors and extensors. Rather, the movement begins as high up as the shoulder blade. During forward movement, the proximal part of the shoulder blade is moved caudally. The shoulder joint swings forward as a result. Now the shoulder joint is extended, causing the upper arm to swing forward. During the standing phase, on the other hand, the proximal part of shoulder blade is pulled in a cranial direction, the shoulder joint moves caudally as a result and the leg swings down. Adduction and abduction movements are possible, in which the chest and trunk musculature are also involved. The most important muscles involved in the movement of the foreleg are presented briefly in the table.

The muscles of the forearm are responsible for flexing and extending the joints below the elbow joint. During movement they are mainly important for stabilising the distal limb and for absorbing the concussion that results when the horse puts its foot down.

The Musculature of the Forehand

Under natural conditions, the horse's forehand essentially makes forwards and backwards movements. The forwards movement is usually described as a swinging or suspended leg phase and the backwards movement is described as the standing phase. The controlled movement is a result of interaction between different muscles.

Did you know …?

The movement in the shoulder blade amounts to around plus/minus two centimetres cranially or caudally. The saddle has to allow the shoulder enough freedom of movement so that the leg can be moved as a whole. If this movement from the shoulder is not present, the horse can only move its front leg from the elbow joint and the forearm. This action can range from small, shuffling movements to a spectacular raising of the knee joint ("knee action") without extension.

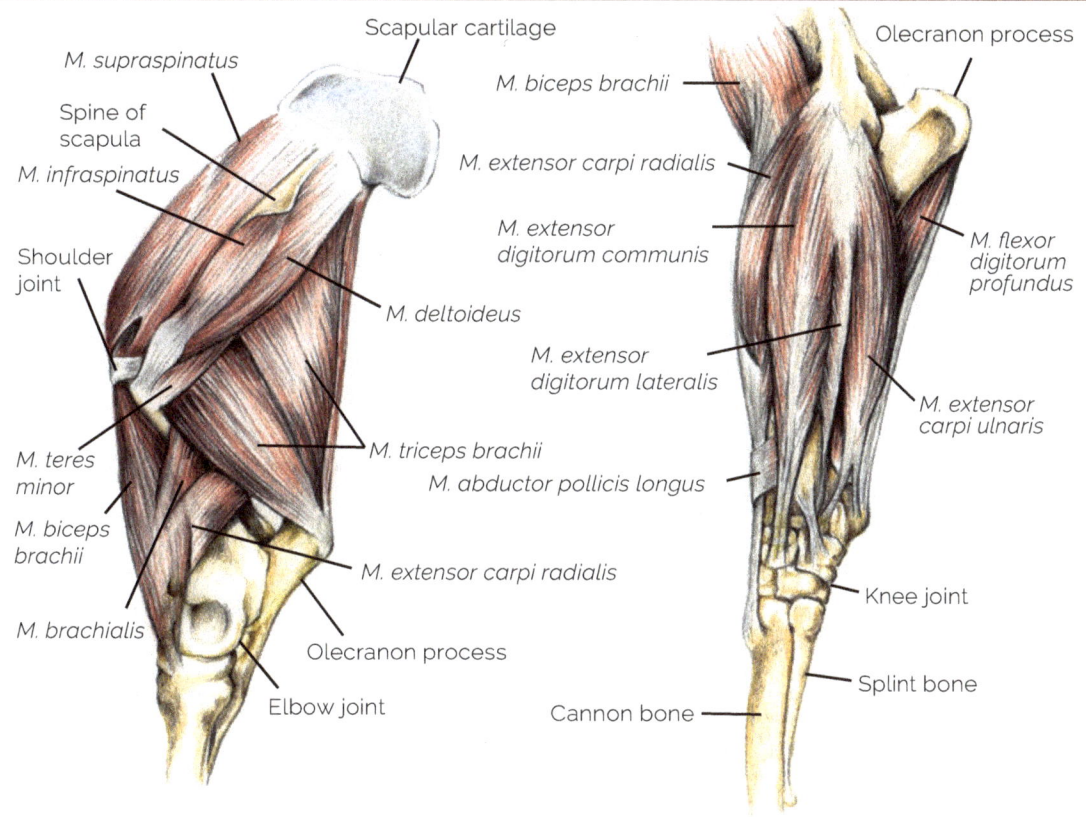

Musculature of the shoulder and upper arm. (Illustration: Retsch-Amschler)

The musculature of the forearm. The flexors and extensors have the important role of stabilising the distal limb and absorbing shock. (Illustration: Retsch-Amschler)

An overview of the essential muscles of the forehand and their roles

Shoulder girdle

M. serratus ventralis thoracis and cervicis — "Serrated muscle girdle", supports the trunk between the shoulder blades, raises the trunk over the standing leg *(thoracic part)* during the standing phase or raises the forehand *(cervicis part)* when the horse is collected.

M. pectoralis ascendens (profundus) and M. subclavius — "Pectoral muscle girdle" supports the trunk between the shoulder blades, pulls the trunk over the standing leg during the standing phase *(M. pectoralis ascendens)*..

The Musculoskeletal System 49

An overview of the essential muscles of the forehand and their roles

M. trapezius
M. rhomboideus
M. latissimus dorsi

Dorsal attachment of the shoulder blade to the trunk, particularly under strain on landing after a jump.

M. pectoralis descendens
M. pectoralis transversus

Ventral attachment of the shoulder blade to the trunk, involved in lateral movement *(M. pectoralis transversus)* and forward movement *(M. pectoralis descendens)* of the foreleg.

Forward movement (protraction)
M. trapezius thoracis
M. serratus ventralis thoracis

Dorsal part of forward movement, the shoulder blade is pulled caudally in the proximal area.

M. pectoralis descendens
M. brachiocephalicus
M. omotransversarius

Ventral part of forward movement, the upper arm and the shoulder joint are pulled cranially and the leg swings forward.

Backward movement (retraction)
M. rhomboideus
M. trapezius cervicis
M. serratus ventralis cervicis

Dorsal part of backward movement, the shoulder blade is pulled cranially in the proximal area, causing the shoulder joint to swing caudally.

M. subclavius
M. pectoralis ascendens (profundus)
M. latissimus dorsi

Ventral part of backward movement, the upper arm is pulled caudally, while the muscles of the shoulder girdle "pull" the trunk over the foreleg.

Adduction
M. pectoralis transversus, descendens and ascendens (profundus)
M. subscapularis

Adduction is mainly achieved using the pectoral muscles. The *M. subscapularis* pulls the shoulder joint to the trunk.

Abduction
M. trapezius
M. rhomboideus
M. infraspinatus
M. deltoideus
M. abductor pollicis longus

The first two muscles pull the proximal area of the shoulder blade to the body so that the distal part, i.e. the shoulder joint can swing outward. The last three muscles then move the leg outward.

The horse shifts all of his weight onto the hindquarters for the levade. (Photo: Slawik)

The Musculature of the Hindquarters

Visually, we tend to consider the croup to be part of the back. However, it is actually part of the hindquarters, along with the pelvis. The musculature of the hindquarters is particularly well pronounced in horses. It is the location of the engine that creates power that is transferred to the forehand via the back muscles and the spinal column. The croup, above all, is especially well pronounced in order to move the horse forward. Correspondingly, we can see defined hindquarters musculature from the side in horses that are well ridden. This musculature is described as looking like "trousers" because it looks like a pair of knee breeches. The heavier a horse is on the forehand, the more weakly developed the musculature of the hindquarters and the more evident the shoulder girdle.

Did you know …?

By looking at *M. serratus ventralis cervicis* and *thoracis* in particular, we can see whether a horse is ridden more over the back or more on the forehand. In a horse that is heavy on the forehand, the muscles will form an "indentation" cranially in front of the shoulder blade and a "bump" caudally behind the shoulder blade. The more the horse is worked through its back, the more this relationship will become inverted. This is because the front part of *M. serratus* "lifts" the shoulder and the base of the neck when weight is shifted onto the hindquarters and only works when the horse sits back. The back part of *M. serratus* "pulls" the trunk over the leg on the ground and mainly works during movement where weight is on the forehand. This muscle is a more reliable indicator of the horse's level of training than the muscles of the back or croup.

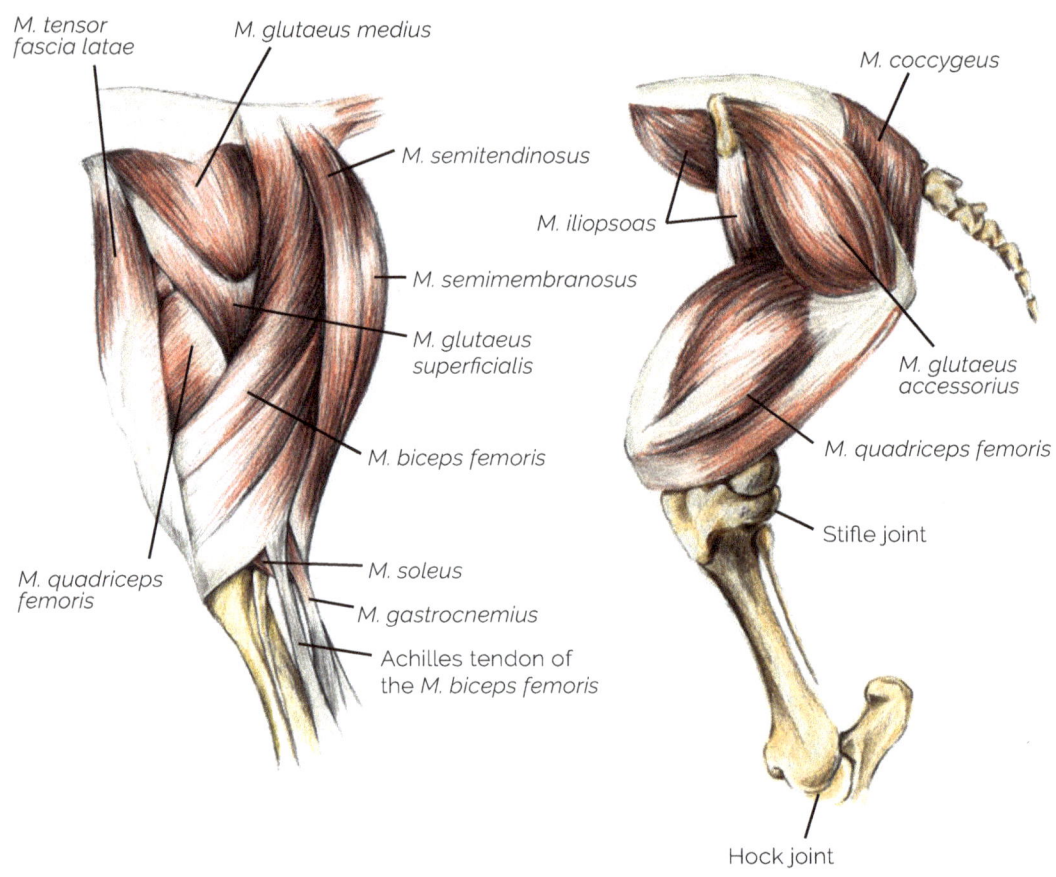

The superficial muscular layer on the thigh and croup. (Illustration: Retsch-Amschler)

The deep muscular layer on the thigh and croup. (Illustration: Retsch-Amschler)

Under natural conditions, the horse's hindquarters essentially makes a forwards and backwards movement, i.e. protraction and retraction. On turns or during lateral movements, abduction and adduction can also be observed. The angles of the hindquarters necessitate much more complicated muscular interplay than in the forehand. For example, *M. quadrizeps femoris*, an extensor of the stifle joint, also requires contraction of its antagonist *M. biceps femoris* in the standing phase, because the stifle would otherwise be overstretched.

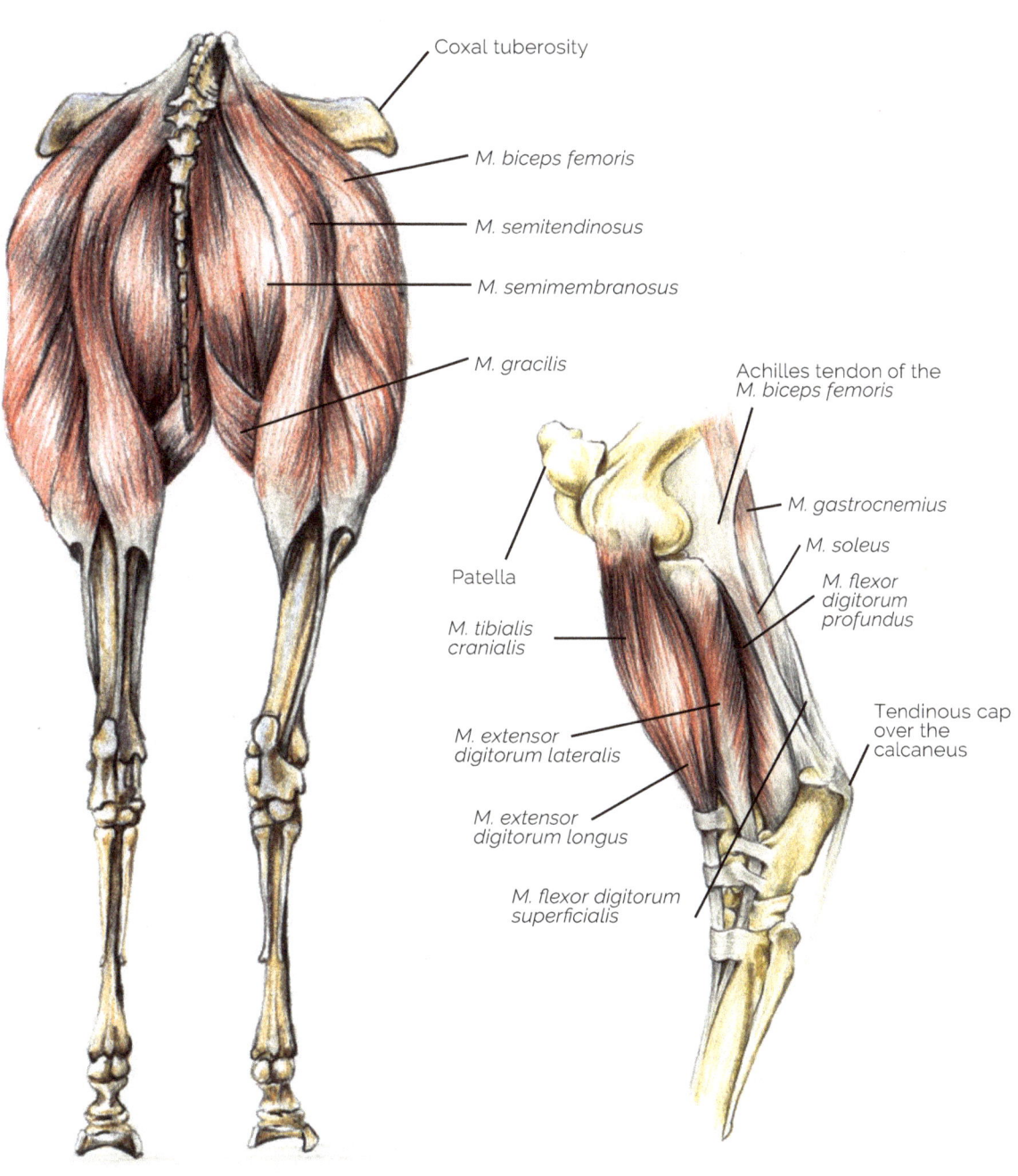

The musculature of the hindquarters viewed caudally. If it is well pronounced, we say that the horse is wearing "trousers".
(Illustration: Retsch-Amschler)

Musculature of the gaskin.

An overview of the essential muscles of the hindquarters and their roles

Musculature of the croup and pelvis
M. glutaeus medius | M. glutaeus superficialis | M. glutaeus accessorius | M. glutaeus profundus
As a whole, these muscles form the croup of the horse, which is visible from the outside. They correspond to the musculature of the buttocks in humans. Their role is primarily to extend the hip joint, but they are also involved in adduction of the hind leg by rotating the hip joint inward. The *M. glutaeus medius* also acts as an extensor of the transition between the lumbar spinal column and the sacrum, as well as the sacroiliac joint (SIJ). The muscles of the pelvic floor are found deep inside the pelvis, but they play a subordinated role in movement. Their role is mainly stabilising the joints and proprioception.

Cranial muscles of the thigh
M. quadriceps femoris | M. tensor fascia latae
When they contract they extend the hip, stifle and hock joints simultaneously and in cooperation.

Caudal muscles of the thigh
M. biceps femoris, cranial part | M. semimembranosus | M. biceps femoris, caudal part | M. semitendinosus
All of these muscles act as extensors of the hip joint and work together with the gluteus musculature in this area. However, because of their different points of influence, they have different effects on the stifle and hock joint. During the standing phase, the cranial part of the *M. biceps femoris* and the *M. semimembranosus* act synergetically with the *M. quadriceps femoris* to extend the knee joint. The caudal part of the *M. biceps femoris* and the *M. semitendinosus*, on the other hand, are antagonists and flex the stifle joint during the standing phase in order to prevent the stifle from overstretching.

Medial muscles of the thigh
M. sartorius, M. gracilis | M. pectineus, Mm. adductores
All of these muscles are arranged on the medial side of the femur and are involved in adduction of the hind leg. They are therefore antagonists to the gluteus musculature.

Cranial muscles of the gaskin
M. tibialis cranialis | M. fibularis tertius (Tendo femorotarseus) |
M. extensor digitorum longus | M. extensor digitorum lateralis
The first two muscles are exclusively involved in flexing the hock joint, while the other muscles are also extensors of the joints in the distal limb.

Caudal muscles of the gaskin
M. gastrocnemius | M. flexor digitorum superficialis | M. popliteus |
M. flexor digitorum profundus | M. tibialis caudalis
The first two muscles are extensors of the hock joint. The role of *M. popliteus* is to flex the stifle and to rotate it inwards. The last two muscles flex the joints of the distal limb. They are all involved in stabilising the fetlock and pastern joint of the hind leg.

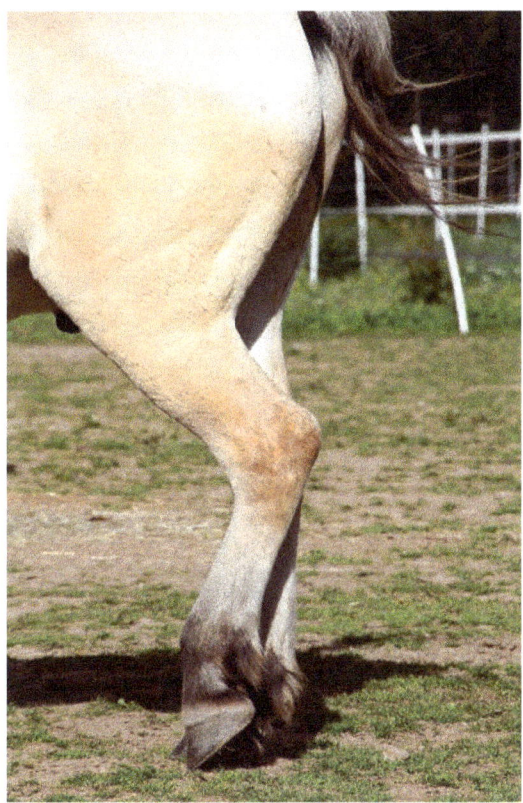

The horse can take the weight off one hind leg when resting if the other hind leg is secured by the bucksaw construction (see diagram opposite). (Photo: Maurer)

Did you know …?

As a flight animal, horses try to avoid lying down for long periods. Nature has developed a "bucksaw" construction in the hind legs to allow the horse to doze whilst standing, without using excessive muscular energy. The muscles and tendons of the hind leg are joined together in such a way that the stifle and hock joint can only be flexed or extended together. The equine patellar tendon has a "loop" that can be hooked over the medial ridge of the femur at the stifle joint. This fixes the stifle joint in its extended position and the hind leg can no longer be bent. The horse can now take the weight off the other hind leg and rest in a standing position, using as little energy as possible.

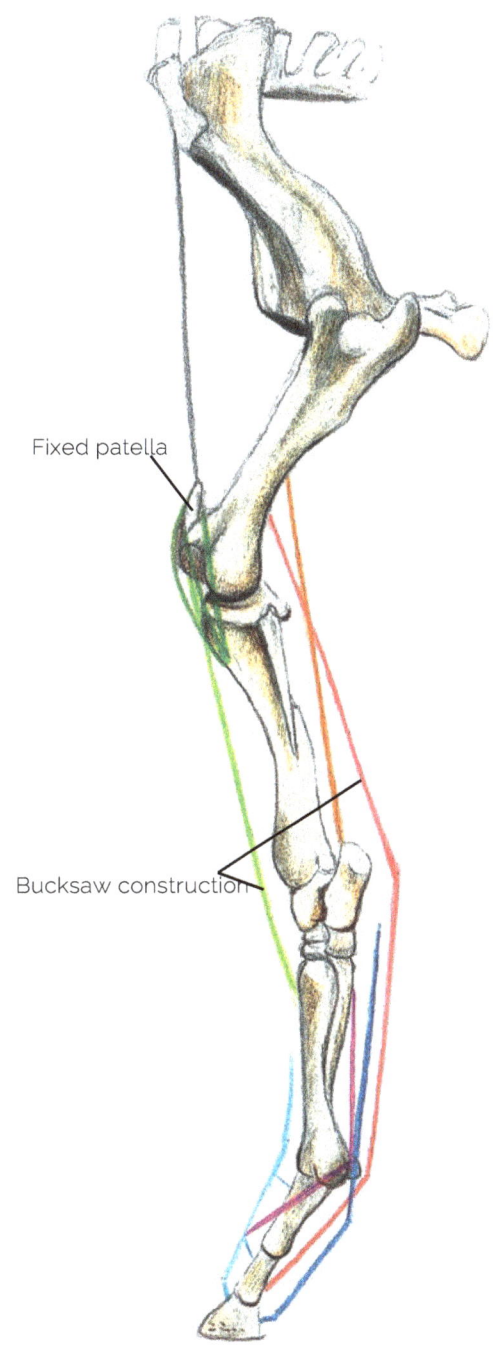

The "bucksaw" construction of the hind leg allows the horse to doze while standing without using excess energy. (Illustration: Denmann)

The Musculoskeletal System

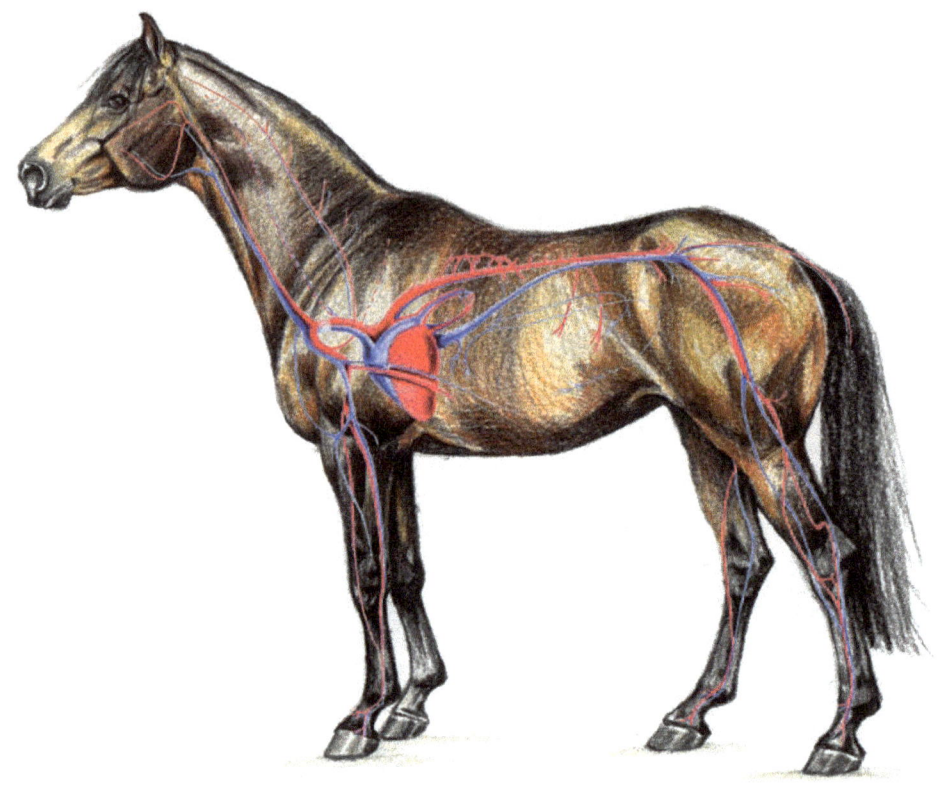

The systemic circulation of the horse. Arteries (vessels that carry oxygenated blood) are shown in red and veins (vessels that carry deoxygenated blood) are shown in blue. (Illustration: Retsch-Amschler)

The Cardiovascular System

Circulation is the basis for everything horses can do. The heart pumps blood through the systemic circulation and through the pulmonary circulation. Blood transports everything that the body's cells need. It carries oxygen and nutrients to the tissue and takes metabolic waste and carbon dioxide with it on the way back. It is also a means of transport for the immune system and for the chemical messengers, the hormones.

A large horse has more than 30 litres of blood. Approximately two-thirds of this blood consists of liquid with dissolved matter and one-third is made of solid constituents. The fluid that remains when the blood clots is called serum. If we separate hirudinised blood using centrifugalisation, we call this fluid plasma. Blood plasma contains many diluted substances, for example, sugar, protein and trace elements.

The Blood Cells

The solid constituents of blood are red blood cells (erythrocytes), white blood cells (leucocytes) and blood platelets (thrombocytes). In a blood count, the solid components can be found under the term "haematocrit".

Erythrocytes are cells with no nucleus that appear red because they contain the pigment haemoglobin. The sole purpose of these cells is to transport oxygen from the lungs into the tissues. Erythrocytes move freely in the blood and are shaped like little plates, but may become distorted as they pass through thinner vessels. As they age they lose this elasticity and can be filtered out by the spleen as a result.

White blood cells, leucocytes, are much less numerous than red blood cells. They are all part of the immune system and there are different versions of them with a very wide range of different purposes. Leucocytes can be divided up into lymphocytes, granulocytes and monocytes. We also differentiate between B and T lymphocytes. Both have a role to play in specific immune defence against germs or toxins because they can form antibodies. Granulocytes are divided up into neutrophile, eosinophile and basophile granulocytes. They also have specific roles in immune defence,

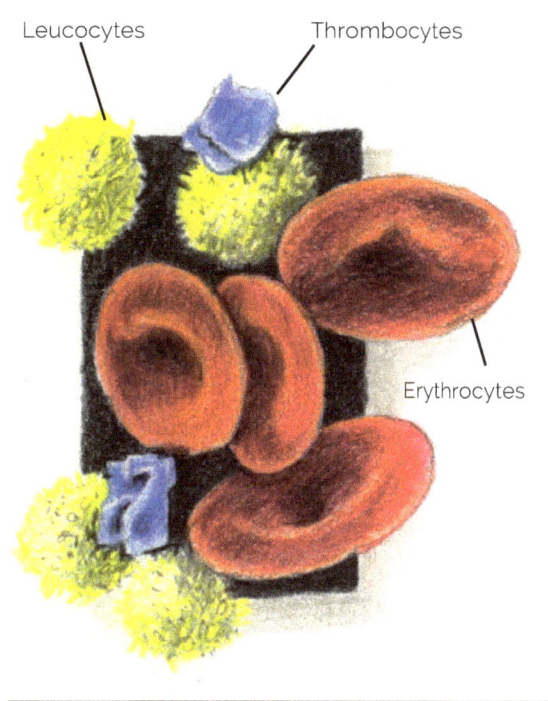

Blood cells. (Illustration: Retsch-Amschler)

for example, an increase in eosinophiles indicates an allergy or parasitic disease. These different types are therefore specified in the differential blood count. Lastly, monocytes are phagocytes that digest any foreign cells or cell detritus.

Blood also contains blood platelets, which are called thrombocytes. They play an important role in blood clotting.

In horses, the spleen is able to store blood cells and add them to the circulation when demand increases. The spleen has several roles. It is a store for blood cells and a filter for old erythrocytes and also contains a series of specialised immune cells that produce antibodies. The equine

Diagram of the blood constituents

	Blood			
Blood cells			**Blood plasma**	
Red blood cells	White blood cells	Blood platelets	Fibrin	Blood serum
Erythrocytes	Leucocytes	Thrombocytes		
Lymphocytes	Granulocytes	Monocytes		
Neutrophile Granulocytes	Eosinophile Granulocytes	Basophile Granulocytes		

Diagram of the blood constituents.

spleen is located in the abdominal cavity, caudally on the left side of the liver. The liver is also a large blood store that can step in if required and pump large quantities of blood into the system.

The Vascular System

In a healthy horse, blood is found in the blood vessels, arteries and veins. Blood in the arteries moves away from the heart. In the systemic circulation, oxygenated blood travels from the heart to the peripheral tissues via the arteries. The journey begins with the principal artery (aorta) that has a shunt into the carotid artery just behind the heart, in order to pump blood in the direction of the head. On its way through the body, the principal artery branches off into various arteries that take blood to the different parts of the body. This is how all of the internal organs and all of the muscles and nerve cells, i.e. every part of the body, is supplied. There is a special feature in the intestines, whereby the blood vessels that run along the intestines then extend into

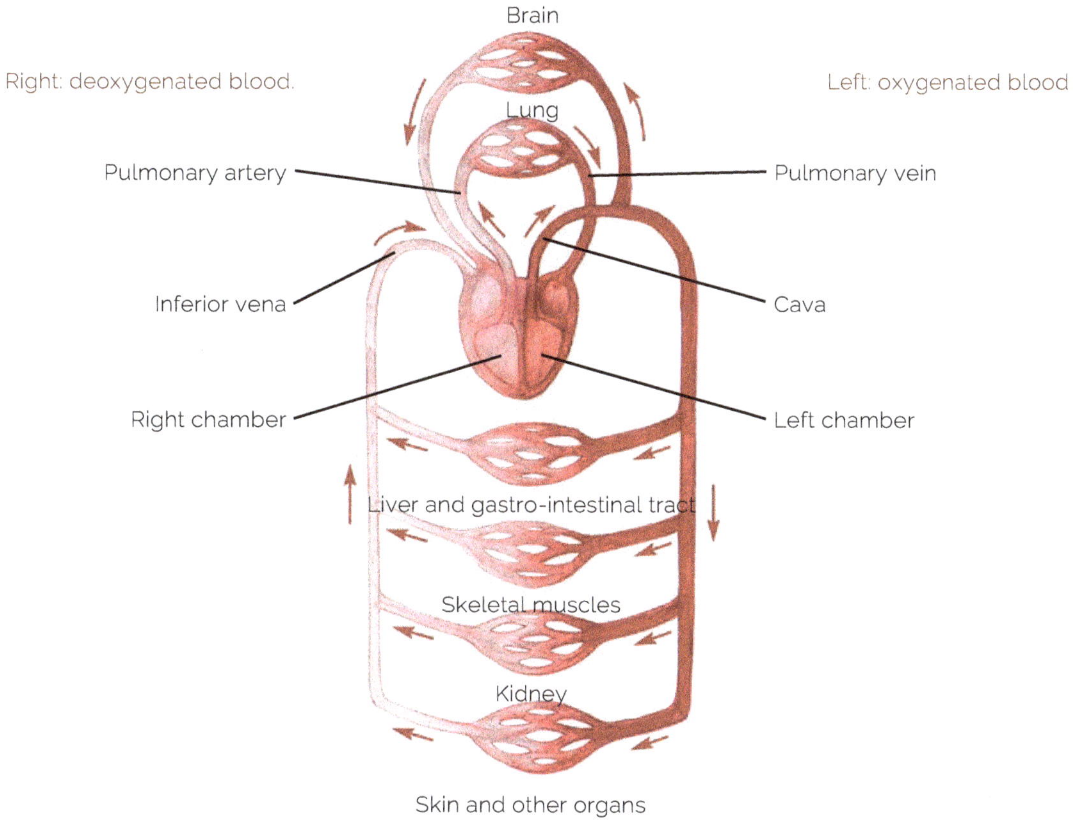

Schematic representation of blood circulation. (Illustration: Mähler)

the liver (portal system). This ensures that all of the substances absorbed from the intestines are sorted and processed in the liver first and that toxins, for example, can be recognised in time and made harmless. Arteries have to withstand high blood pressure and have strong and distinctly muscled walls.

We can feel a pulse in the arteries. In horses, a pulse can be felt on the inside of the lower jaw and on the underside of the dock. The arteries branch out and become many, increasingly thin vessels called arterioles. The arterioles branch out again and form a capillary bed in the tissue made up of extremely fine blood

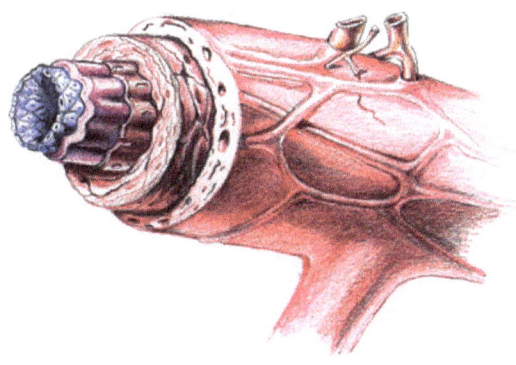

Arteries have strong walls so that they can withstand high blood pressure. (Illustration: Retsch-Amschler)

vessels. The exchange with the environment takes place through the walls of these vessels. Oxygen and nutrients are absorbed here and carbon dioxide is released. The pressure in the capillary bed is so high that liquid from the blood vessels floods into the surrounding tissue. From this point in time, the fluid is called lymph and not plasma. The lymph washes around the cells, brings them nutrients from the blood stream and takes away the cells' waste products on its way. Later it collects in the lymph tracts and is added back to the venous circulation via the lymphatic system. The capillaries change into very thin veins in which the now deoxygenated blood is returned to the heart in increasingly wide channels.

But why does the blood flow back to the heart? The heart actively pumps blood into the arteries under pressure, but it cannot draw the blood back again. Therefore, there are other mechanisms that transport venous blood back to the heart. The horse has a frog (see also chapter "Structure of the Distal Limb and Hoof") to stop the blood from pooling in the legs. The frog moves the blood from the hoof back into the area where the musculature is found. There the veins run between the skeletal muscles. Each muscle contraction exerts pressure on the veins and when the muscle relaxes, the vein relaxes, which also creates a pump action for transporting venous blood. Veins are fitted with valves that only open in the direction of flow to prevent the blood from flowing back into the legs. Constant, quiet movement is therefore necessary for good circulation. If the horse stands for a long time, in the stable for example, congestion takes place in the legs, a condition that we call "filled legs".

Did you know ...?

There is a special feature in the blood circulation in the horse's legs that is called anastomosis. It involves "shunts" between the arteries and veins that can be opened and closed as required. When they are closed, all of the arterial blood gets right down into the hoof area, from which it is then transported back in the direction of the heart by the venous system. If the anastomoses are open, a large portion of arterial blood flows into the venous system and straight back into the body at the knee or hock joint. This minimises blood circulation in the legs. This mechanism is necessary in the cold, for example, because the blood would otherwise cool down too much in the legs. It is also used in states of shock or stress, i.e. when more blood is needed in the body. In these cases, horses will have cold feet.

Pulmonary circulation is different from systemic circulation, because the two large pulmonary veins that lead to the heart carry bright red, oxygeated blood, unlike the systemic circulation where the venous blood is deoxygenated and therefore dark red. This happens because deoxygenated blood enriched with carbon dioxide flows from the body into the heart. It gets into the pulmonary circulation through the right half of the heart, so it is initially pumped to the lungs from the heart. The lung is made up of thousands of little air cells called alveoli. Each alveolus is surrounded by a fine capillary net. The exchange of gases takes place here. Carbon dioxide is released in the lungs by the blood and oxygen from the lungs is bound with the haemoglobin of the

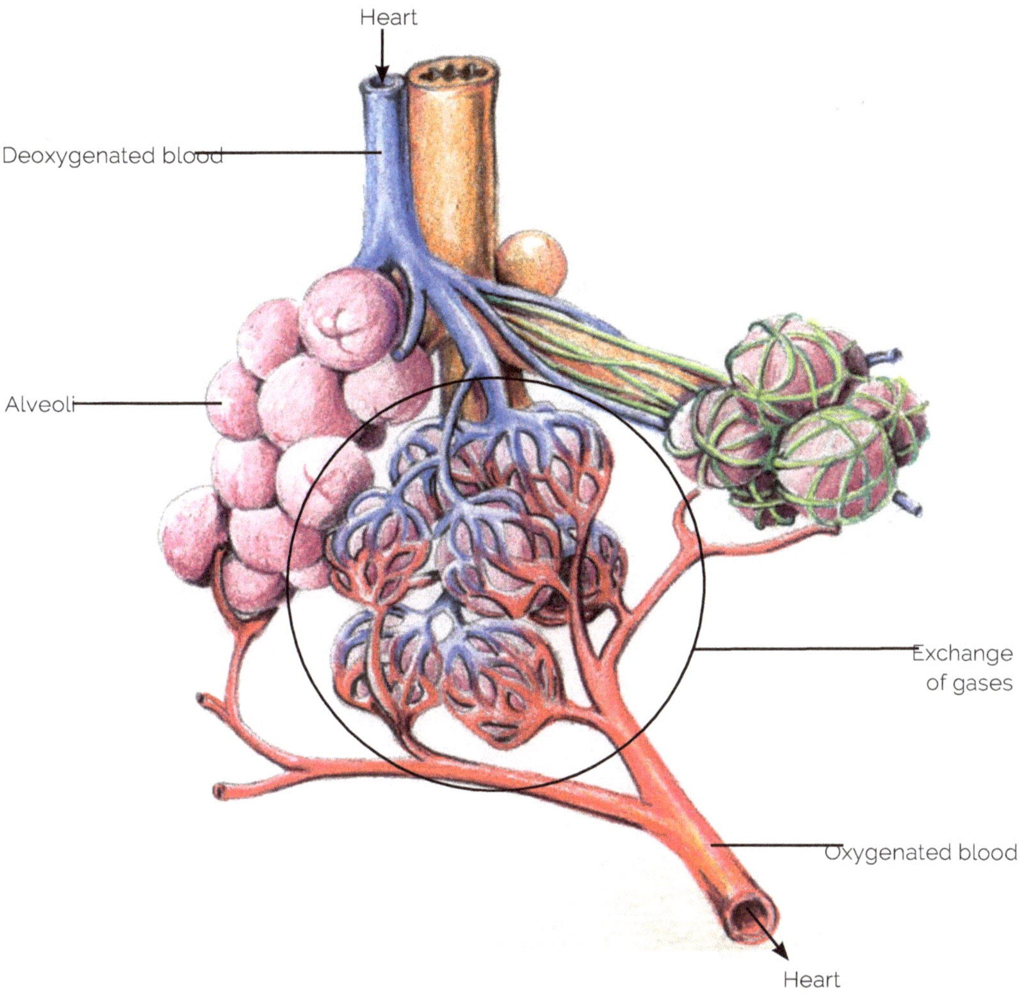

The exchange of gases takes place in the capillary network around the alveoli, so that oxygenated blood can flow back to the heart. (Illustration: Retsch-Amschler)

erythrocytes. The pulmonary veins now transport oxygenated, bright red blood back to the left half of the heart. From here it is pumped into the body again. The heart diverts off some of this oxygenated blood for its own purposes and sends it to the coronary vessels in order to supply its own muscle cells with oxygen.

The Structure of the Heart

The horse's heart lies cranial–ventral in the thorax, precisely between the forelegs behind the sternum. It can be heard on the left side between the olecranon process and the thorax and sometimes its beat can even be felt with your hand. The heart works around the clock and cannot be voluntarily

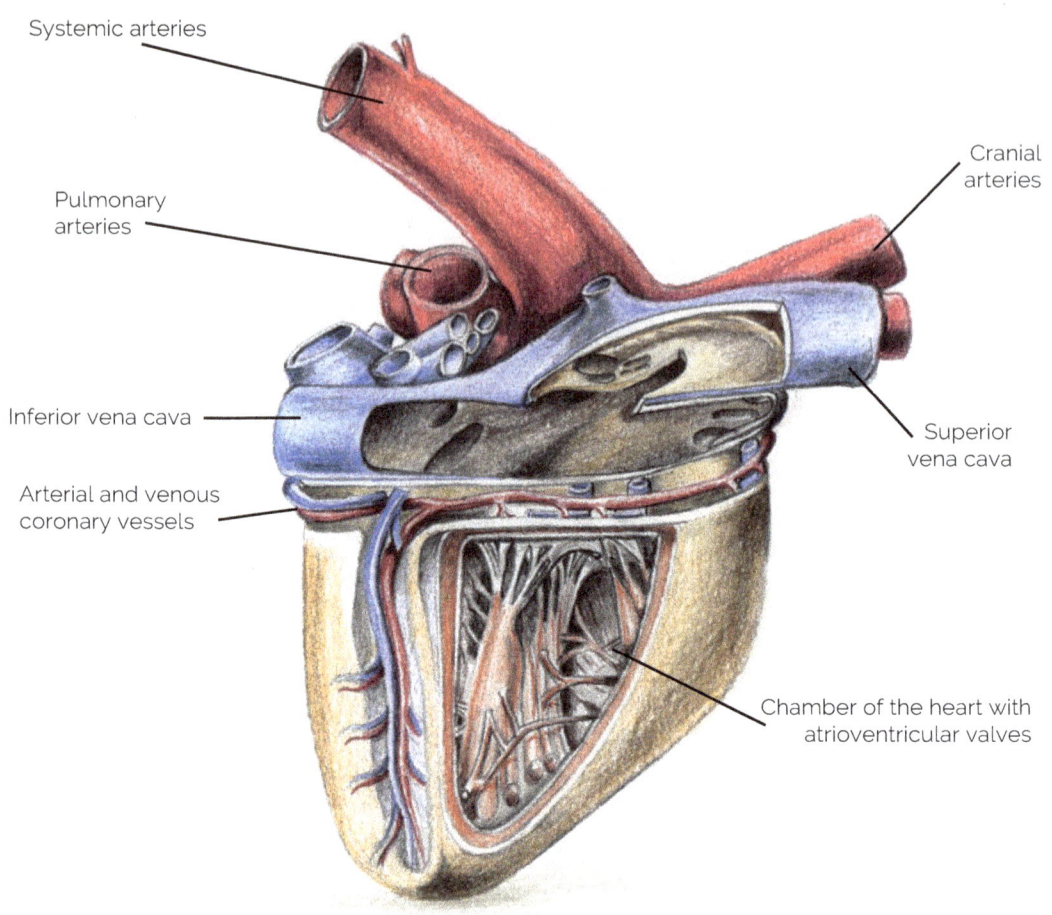

The structure of the heart. (Illustration: Retsch-Amschler)

controlled. Depending on the weight and breed of horse, an equine heart weighs between 0.6 and 1 percent of its body mass, i.e. around 5 kilos in a large horse. The heart is protected from the outside by the pericardium, a kind of double-walled bag made from firm connective tissue. This large hollow muscle is the engine of the entire body. It consists of a left and a right half that are completely separated from one another by a wall. Each half has two chambers, an atrium and a main chamber that are joined together by specially designed valves. The heart is larger and has thicker walls on the left than on the right, because the left half of the heart has to do more work. The left heart chamber contracts to pump blood into the aorta through a kind of hinged door. This hinged door consists of three semilunar valves that close after the heart has contracted and prevent the blood from flowing back in. A secure closure is necessary so that enough pressure can be built up during the contraction before the valves open. When these valves open, the blood is propelled into the body, creating blood pressure.

At the same time, the semilunar valves open the right main chamber and the deoxygenated blood collected there is pumped into the lungs. The blood reaches the chambers through the atria. The left atrium gets oxygenated blood from the pulmonary veins. The right atrium gets deoxygenated blood from the body. The atria are joined to the chambers of the heart by a clever valve system. The left atrium contracts while the left chamber is relaxed. As this happens, the blood flows from the left atrium into the left chamber through a bicuspid atrioventricular valve, which roughly resembles a stretched-out awning and is held in place by powerful muscles. At the same time, blood from the atrium flows into the right half of the heart through a tricuspid atrioventricular valve in the right chamber of the heart. If the main chambers are full they contract and the blood is pumped out of the heart again, the atria can refill and the cycle starts again from the beginning.

Did you know …?

Unborn foals have a hole in the septum of the heart so that blood can flow between the right and left halves of the heart. This opening closes shortly after birth. If the hole does not close ("hole in the heart"), the horse will never be able to perform well, because it cannot adequately enrich the blood for systemic circulation with oxygen because of constant mixing.

Did you know …?

The sounds that are made when the valves close can be heard with a stethoscope, mainly on the left side of the thorax at the sternum. We call these heart sounds. If sounds are present other than those of the valves closing, we can assume that the heart is not completely healthy. Heart defects are very common. There are studies that show that around 40 percent of apparently healthy horses have a heart defect. The defect is often left-sided cardiac insufficiency, especially in old horses and sport horses.

At rest, a horse's heart beats approximately 36 times a minute, i.e. more than 50,000 times a day. When a horse is working and the body needs more oxygen, the heart can beat much faster and under high performance it can perform more than 100 contractions a minute. At the same time, it can increase its cardiac output so that more blood is pumped through per contraction. This volume can principally be increased by training.

The heart is predominantly self-determined, unlike the skeletal musculature, which is controlled by the brain. The sinoatrial node, a pacemaker for the heart beats, is located in the wall of the right atrium. The stimulation created there jumps from one cardiac muscle to the next via the gap junctions, resulting in a peristaltic contraction of the entire cardiac muscle. However, the heart rate is controlled by the autonomic nervous system, which, for example, releases the hormone adrenaline in preparation for flight, increasing the heart rate so that the skeletal muscles are as well supplied as possible.

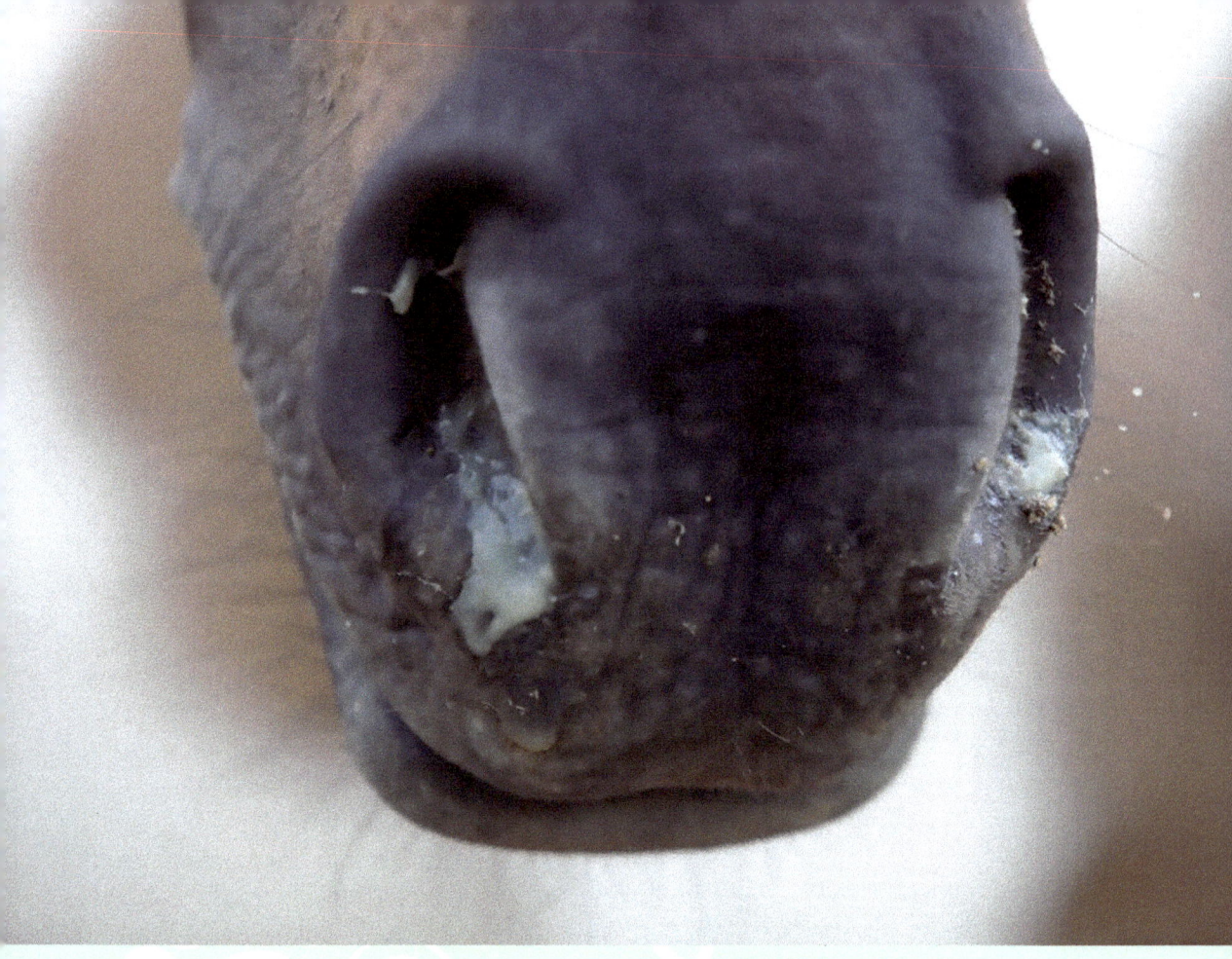

The lymphatic system is often very busy if the horse has a cold. (Photo: Slawik)

The Lymphatic System

The lymphatic system is the body's rubbish collection service. Waste products from cell metabolism are added to the lymph as it flows through the tissue past the cells. The lymph then collects in the lymph tracts, which are interspersed at intervals by filter stations called lymph nodes. Cell detritus, bacteria and silar waste is filtered out and disposed of here. Liquid waste is transported further with the lymph. Just before the heart, the lymphatic system leads to the anterior vena cava so that the liquid can be added back into the blood. Waste is carried by the blood system to the detoxification and excretory organs: liver, kidneys, lungs, skin.

At the same time, the lymphatic system also plays an important role in immune defence. The lymph nodes contain macrophages that belong to the white blood cells (leucocytes). They are phagocytes, i.e. endogenic cells that are able to eat and digest other cells such as bacteria or cell detritus. These phagocytes are swept along through the whole body with the lymph and can also actively immigrate into areas where infections by pathogens have occurred, for example, as a result of injury. They then fight the foreign cells where they find them. In cases of disease, we can observe swelling of the very active lymph nodes, which are filled with macrophages. The lymph nodes of the mandibular space swell so much especially during colds that they can easily be felt.

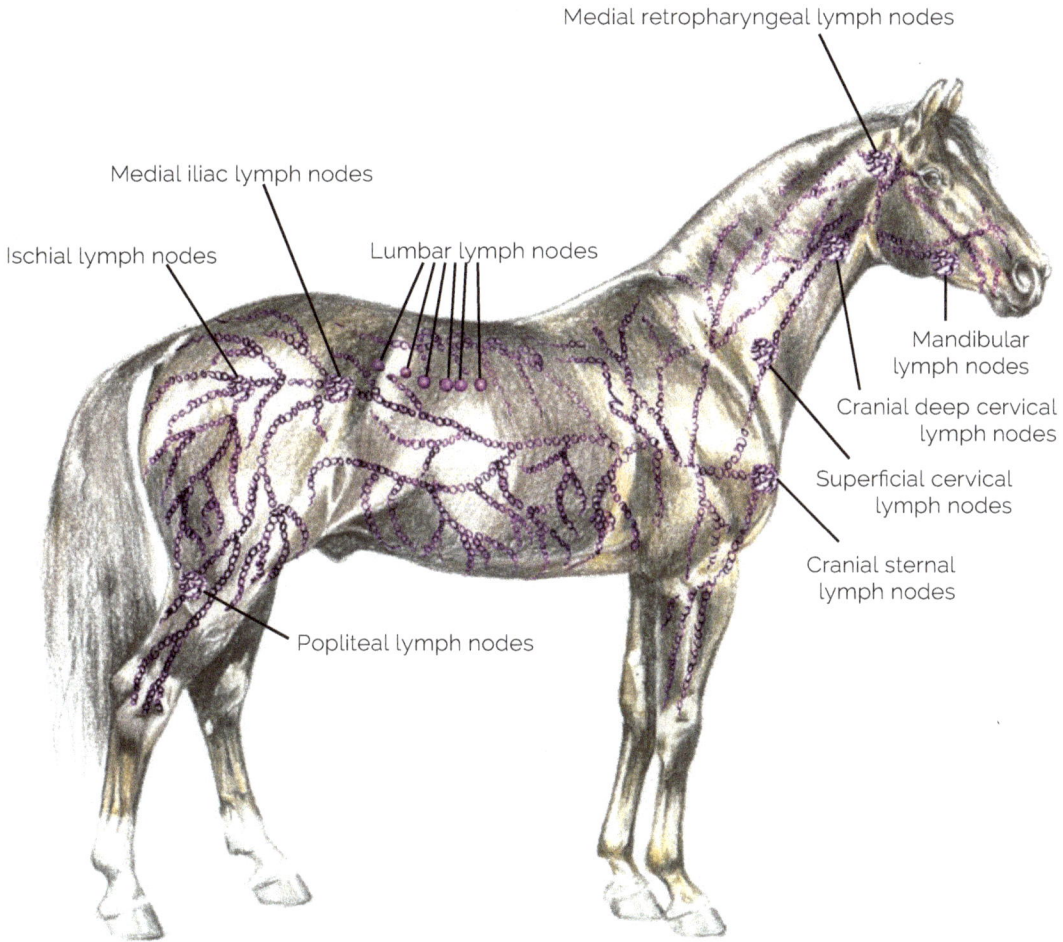

The lymphatic system of the horse. (Illustration: Retsch-Amschler)

Unlike people, horses can only breathe through their noses. (Photo: Slawik)

The Respiratory System

The respiratory system starts at the nostrils where respiratory air flows into the skull. Unlike dogs or human beings, horses can only breathe through their noses and not through their mouths. The respiratory air then flows through the paranasal sinuses of the skull, where it is warmed and moistened and dust is filtered out.

The ethmoid bone is located under the forehead. It is a spongy bone that is covered by mucous membrane. Dust and invading pathogens are filtered at the ethmoid bone, which is why respiratory tract diseases often establish themselves in the upper airways first and not in the lungs. Horses can suffer from runny noses and painful paranasal sinuses too! The mucous membrane of the ethmoid bone can form more mucous that then flows out of the nostrils or is coughed up, removing dirt and bacteria from the airways. Furthermore, the olfactory receptors are located in the mucous membrane of the ethmoid bone, allowing the horse to perceive odours in his environment at the same time as breathing in. This is covered in more detail in the chapter "The Horse's Senses". The larynx, which has one opening to the oesophagus and one to the trachea, is located in the pharynx.

The epiglottis closes the windpipe (trachea) when the horse swallows. It is open when the horse breathes and the respiratory air flows through the upper part of the pharynx into the trachea, a kind of tube that is stretched over cartilage bridges so that it is always open. The inside of the trachea is lined with ciliated epithelium. This is a mucous membrane covered with fine hairs that make flickering movements in the direction of the exit. If dust or food is accidentally carried into the trachea, it sticks to the mucous membrane and is transported back by the hairs. Furthermore, a coughing reflex is triggered to carry these solid objects out of the airways more quickly.

Above the heart, the trachea divides into two primary bronchi that branch off to the left and right. They then keep on branching out into ever finer bronchioles that finally end in alveoli that are surrounded by fine capillary nets. This is where the exchange of gases between oxygen and carbon dioxide takes place.

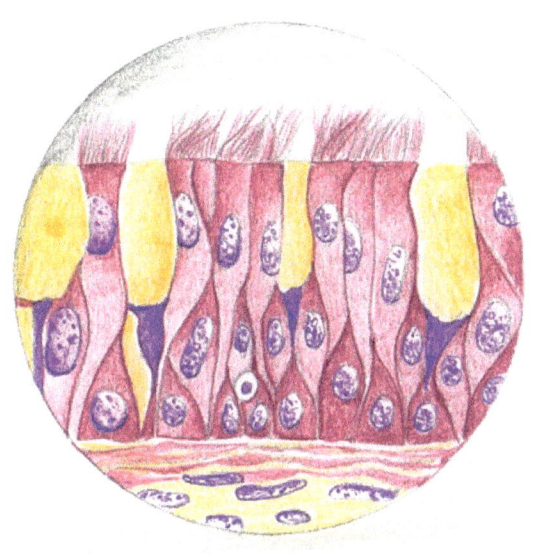

A look inside the windpipe: the cilia (at the top of the picture) transport foreign bodies such as dust or feed outwards. (Illustration: Retsch-Amschler)

Muscular action is needed to fill the lungs. When the horse breathes in, the diaphragm, which is stretched in a dome shape between the abdominal and thoracic cavities, tenses and flattens out. As a result, the intestines are pushed away and the abdomen becomes "fatter" in the flank area. The flattening of the diaphragm creates negative pressure in the thoracic cavity and the air flows in. When the diaphragm relaxes it becomes a dome again and

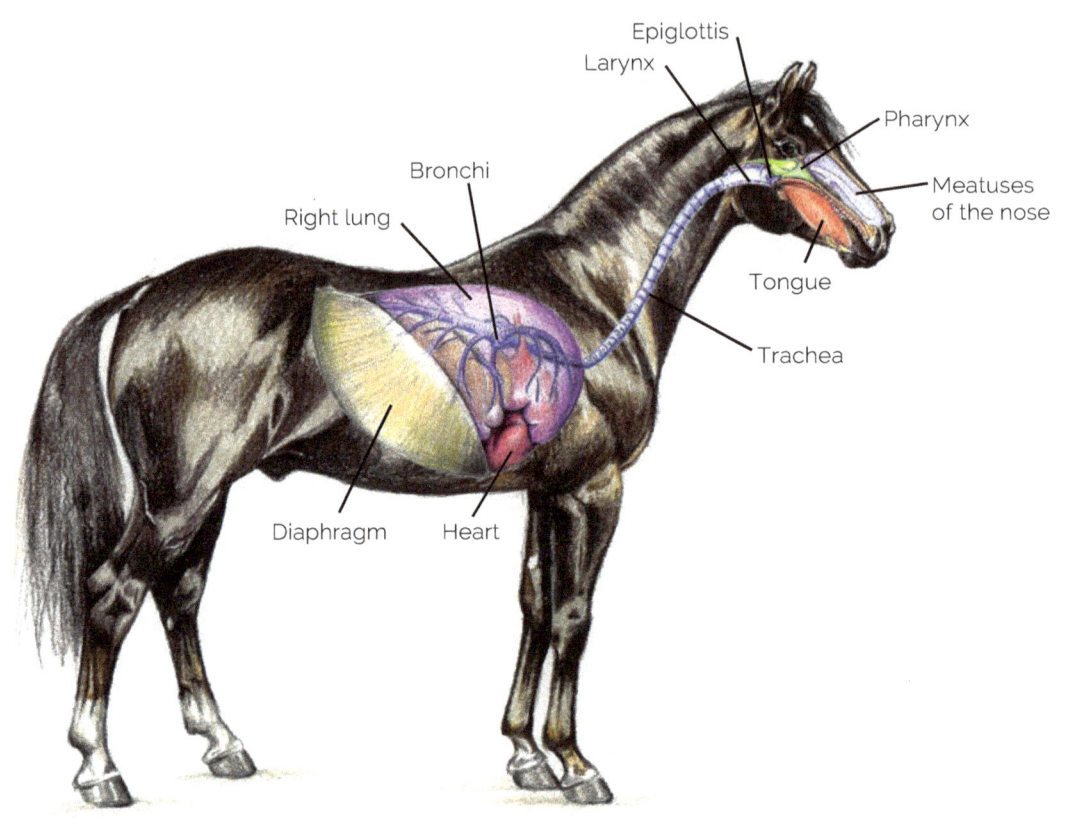

The respiratory system of the horse. (Illustration: Retsch-Amschler)

pushes the air out of the lungs, making the abdomen appear "thinner". As well as abdominal breathing, there is also thoracic or costal breathing. The intercostal muscles are found between the ribs. When these muscles contract, the thorax expands. This also creates negative pressure in the thoracic cavity and the air can flow in. When the intercostal muscles relax, the ribs tip into their normal position and squeeze the air out of the lungs. Breathing in is therefore an active process where the muscles have to work. In a healthy horse, breathing out is a passive process.

Did you know …?

In horses with recurrent airway obstruction ("broken winded"), most of the alveoli are overstretched. As a result, breathing out completely is no longer possible. In order to be able to still squeeze air out of the lungs, these horses use "double lift" breathing. First the air flows out passively because the diaphragm and the intercostal muscles relax. In the second step, the abdominal musculature is tensed and the air is actively pressed out of the lungs. As a result, clearly pronounced abdominal musculature appears over time, which is known as a heave line.

The horse's natural way of feeding is grazing. (Photo: Slawik)

The Digestive System

The horse's digestive system starts at the head, or rather, at the mouth. The lips are very muscular and tear long grass, which is then carried further towards the mouth by the lips and tongue. Horses can also bite off short grass, tree bark and similar vegetation with their incisors. The tongue is a powerful and very flexible muscle that moves food back and forth between the molars. It sorts out unpleasantly tasting or unhealthy plants and they fall out of the side of the mouth.

The teeth grind up roughage that the horse has picked up. Chewing causes enormous quantities of saliva, up to five litres per kilo of hay, to be produced by the salivary glands located under the ears and between the ramus of the lower jaw.

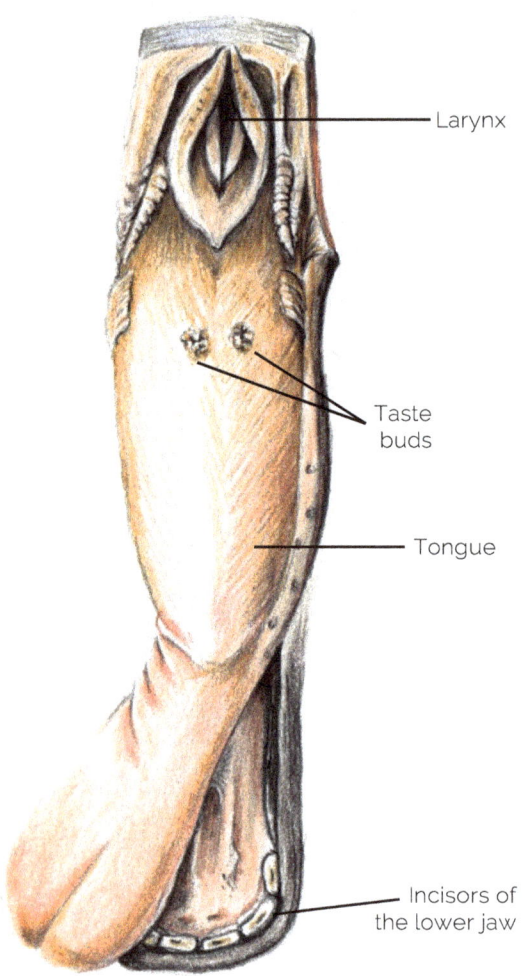

The horse can use its flexible tongue to sift out bad tasting plants or foreign bodies that have been picked up accidentally.
(Illustration: Retsch-Amschler)

The chewed up food is then swallowed through the larynx. The tongue is drawn back during swallowing, the epiglottis is pressed onto the entrance to the trachea and the food pulp flows through the lower part of the pharynx into the oesophagus.

The oesophagus, also known as the gullet, is a muscular tube that, unlike the windpipe, does not contain any cartilage ridges. The smooth muscles of the oesophagus make a peristaltic wave movement and transport the bolus in the direction of the stomach. The horse's oesophagus is very long. It has to reach down the entire neck, through the whole of the thoracic cavity, through a hole in the diaphragm and into the stomach, which lies behind it in the abdominal cavity. The entrance to the stomach is closed by a very solid muscle *(M. sphincter cardiae)*, which, together with the anatomy of the stomach, means that horses cannot vomit. Therefore, once something is in the stomach, it has to go through a total of around 30 to 40 metres of intestine, before it comes out again.

The Stomach

The horse's stomach is like a bag and it is relatively small in comparison with the size of the horse. It has an entrance *(cardia)* at one end and an exit *(pylorus)* at the other. The stomach is made up of two areas. The front area is also called saccus caecus. It is slightly acidic because it is colonised by lactic acid bacteria that predigest various carbohydrates and proteins. The rear area *(pars glandularis)* has many glands that produce gastric juice and gastric acid *(muriatic acid)*, so it is very acidic. The mucous membrane in this area is

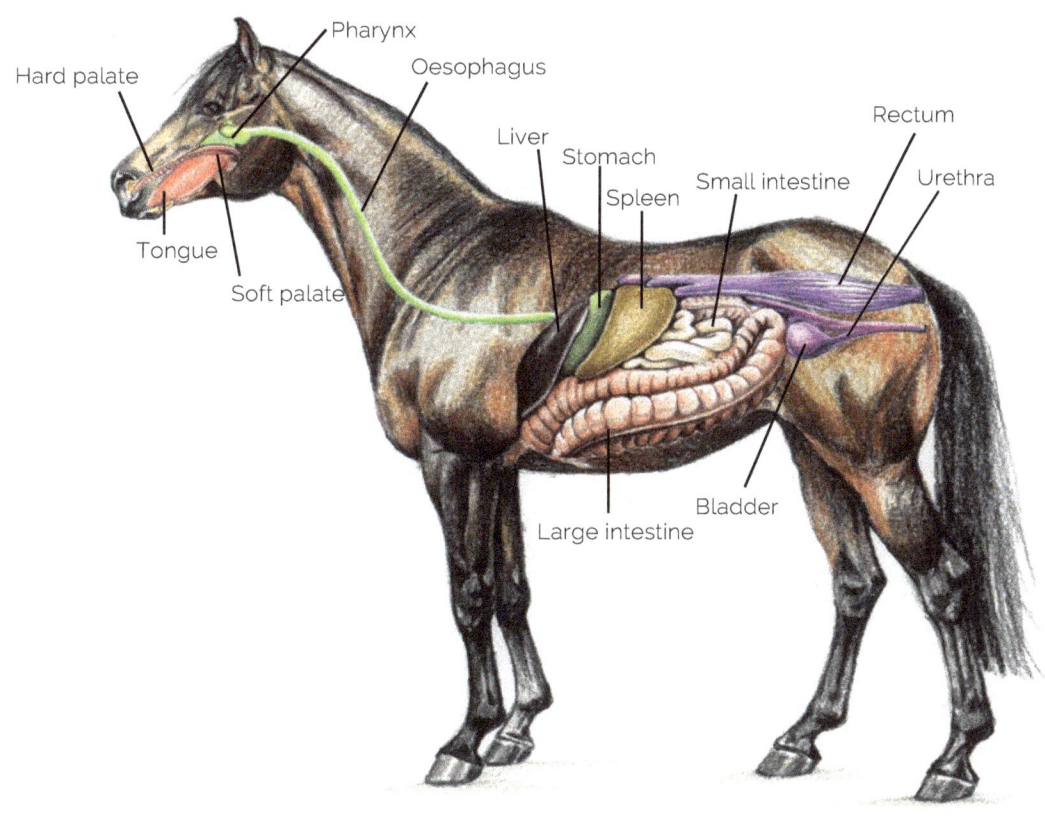

The horse's digestive tract viewed from the left. (Illustration: Retsch-Amschler)

correspondingly also protected against acid, while the front area is very sensitive to gastric acid. The horse's stomach constantly produces gastric acid, whether or not the horse eats. If the horse does not eat any roughage for longer than four hours, a pool of hydrochloric acid results that is "sloshed around" by movement and can cause gastric ulcers in the saccus caecus.

The Small Intestine

In horses, the intestinal tract takes up the majority of their abdominal cavity. The stomach is followed by the small intestine. The small intestine is divided up into three sections: the duodenum, the jejunum and the ileum. The entire small intestine is approximately 20 to 30 metres long and holds 40 to 50 litres.

The acidic bolus is neutralised as soon as it passes out of the stomach through the pylorus. To neutralise the bolus, the horse uses bile, which is produced by the liver and constantly released into the small intestine. Secretions from the pancreas also have a buffering effect.

Bile also contains substances that are able to emulsify fats, i.e. make them water soluble. Without this it would not be possible to digest fat. The pancreatic secretion also contains a number of different digestive enzymes. These are chemical substances that are able to break down nutrients. The particles can then be absorbed through the very permeable wall of the small intestine and transported into the liver with the blood via the hepatic portal vein. In order to extract as many nutrients as possible from the bolus, the mucous membrane is not only very permeable but has lots of little "fingers" (intestinal villi) that reach into the inside of the small intestine. The villi make the surface area of the small intestine 500 times larger! Smooth muscles in the intestinal wall ensure that the bolus is not only transported forwards but also that it is rotated so that new parts of the bolus are always in contact with the intestinal wall, allowing nutrients to be extracted. After 1.5 to 2 hours, the bolus leaves the small intestine through the ileocaecal valve and enters the large intestine.

The Large Intestine

The first section of the large intestine, which connects to the small intestine, is the *caecum*. In horses, it is around a metre long and holds more than 30 litres. It begins at the right inguen and then loops once over the back so that its apex is on the left side between the last rib and the coxal tuberosity. This is also where vets listen in cases of colic, in order to check whether the bolus is still moving forward in the caecum. The wall of the caecum is traversed by a complex arrangement of smooth muscle cells for transporting the bolus. Food is digested bacterially in the caecum. No mammals are able to enzymatically digest cellulose, the material used to make plant cells, in the small intestine. Only micro-organisms can crack these complex molecules and the make the energy that they contain available to mammals. Foals eat their mother's dung in order to absorb the micro-organisms it contains so that they colonize the foal's small intestine. Horses have a symbiotic relationship with these micro-organisms, which means that they help each other. The horse keeps the micro-organisms constantly supplied with food, warmth and moisture and, in return, the micro-organisms provide the horse with the building blocks of cellulose as energy sources, as well as B and K

Did you know ...?

The ileocaecal valve ensures that the bolus can only move in one direction, namely from the small intestine into the large intestine. This is essential, because the large intestine is populated by many micro-organisms, such as bacteria, fungi and protozoa. If these micro-organisms got into the small intestine, there would be nothing to prevent their waste products and toxins from getting into the horse's blood circulation.

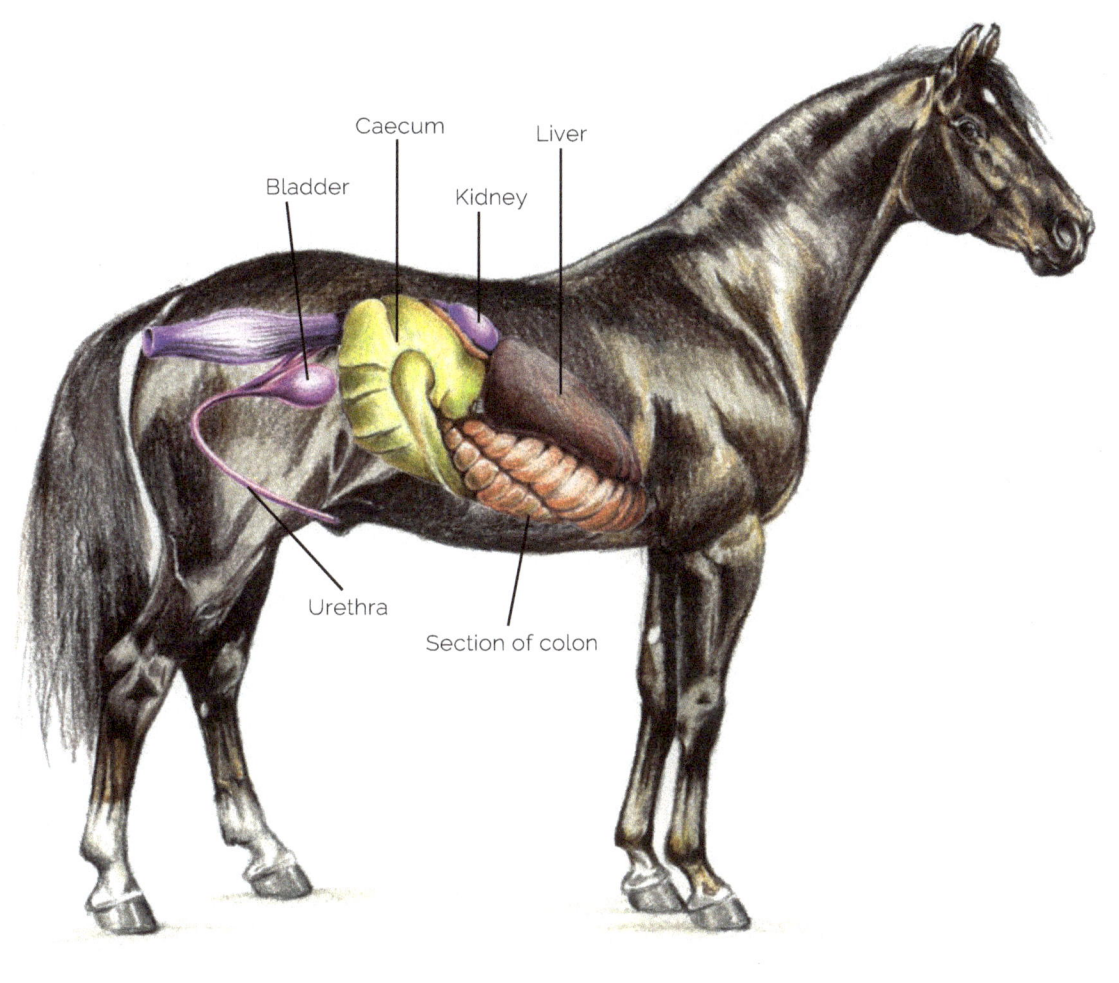

The horse's digestive tract viewed from the right. (Illustration: Retsch-Amschler)

vitamins and some essential amino acids. The process of bacterial breakdown of cellulose is called fermentation. It takes a relatively long time, which is why the bolus moves forward considerably more slowly in the large intestine than in the small intestine.

The food moves from the caecum into the colon. In human beings and dogs, the colon is very short and is mainly important for recovering water from food. At up to 10 metres long and with a capacity of 50 to 60 litres, it is comparatively enormous in horses, because they mainly get their energy from cellulose. In order to fit into the abdominal cavity, it lies in a horseshoe shape on the ventral side in the abdominal cavity, starting in the right ventral inguen. Then, at the left inguen, it takes a turn in the direction of

The Digestive System 73

the back and then lies in a horseshoe shape again on the dorsal side in the abdominal cavity, before ending in the rectum. These many turns make the large intestine very prone to blockages. Furthermore, it has a stomach-like widening in the area of the right kidney in which sand can easily be deposited, which can then lead to dangerous sand colic. Excrement remains in the rectum until it is expelled and the balls of dung are formed. Excretion takes place through the anus, which is located under the dock, approximately 24 to 65 hours after eating.

The Liver

The horse's liver *(hepar)* is located in the abdominal cavity, right next to the diaphragm. This means that the liver is constantly "massaged" by the diaphragm, which flattens and bulges as the horse breathes, ensuring good blood circulation. The liver is the body's large chemistry laboratory. Its tasks include:

- Secretion of bile
- Involvement in carbohydrate metabolism
- Involvement in fat metabolism
- Involvement in protein metabolism
- Storage of various vitamins and trace elements
- Involvement in regulating the hormone balance
- Detoxification function for metabolic waste and toxins that are absorbed
- Blood reservoir (also formation of blood cells before birth) and breakdown of haemoglobin
- Involvement in regulating the water balance
- Heat is released as a "by-product" of the intense metabolic activity and this heat makes up a considerable amount of the body's core heat.

The ability of the horse's liver to detoxify is rather poor in comparison with a dog's, because horses do not frequently come into contact with rotten food in the wild. As a result, the liver is all the better at making glucose from the building blocks of cellulose and primarily making it available to the musculature as a source of energy.

The Pancreas

The *pancreas* is made up of two parts - endocrine and exocrine. We dealt with the exocrine part when we were looking at the small intestine. It contains the digestive enzymes for carbohydrates, proteins and fats and neutralisers for the acidic bolus. The endocrine part produces hormones that control blood sugar levels. The best known hormone, insulin, is released when blood sugar levels increase because of a sugary meal. It ensures that the liver filters out and temporarily stores more sugar from the blood. The insulin's antagonist is called glucagon and is also produced in the pancreas. It ensures that more sugar is released from the liver when blood sugar levels fall, increasing the blood sugar level again.

Contact between stallion and mare prior to mating. (Photo: Slawik)

The Genitourinary System

In anatomy, the urinary organs and the reproductive organs are considered together as the genitourinary system, because they share the same structures. For example, in stallions the urethra is where urine exits the body, as well as the exit for semen during sexual intercourse.

Kidneys and Bladder

Let's begin with the urinary tract. The body has several different ways of disposing of waste. Indigestible constituents of food, as well as some waste products from the liver that are

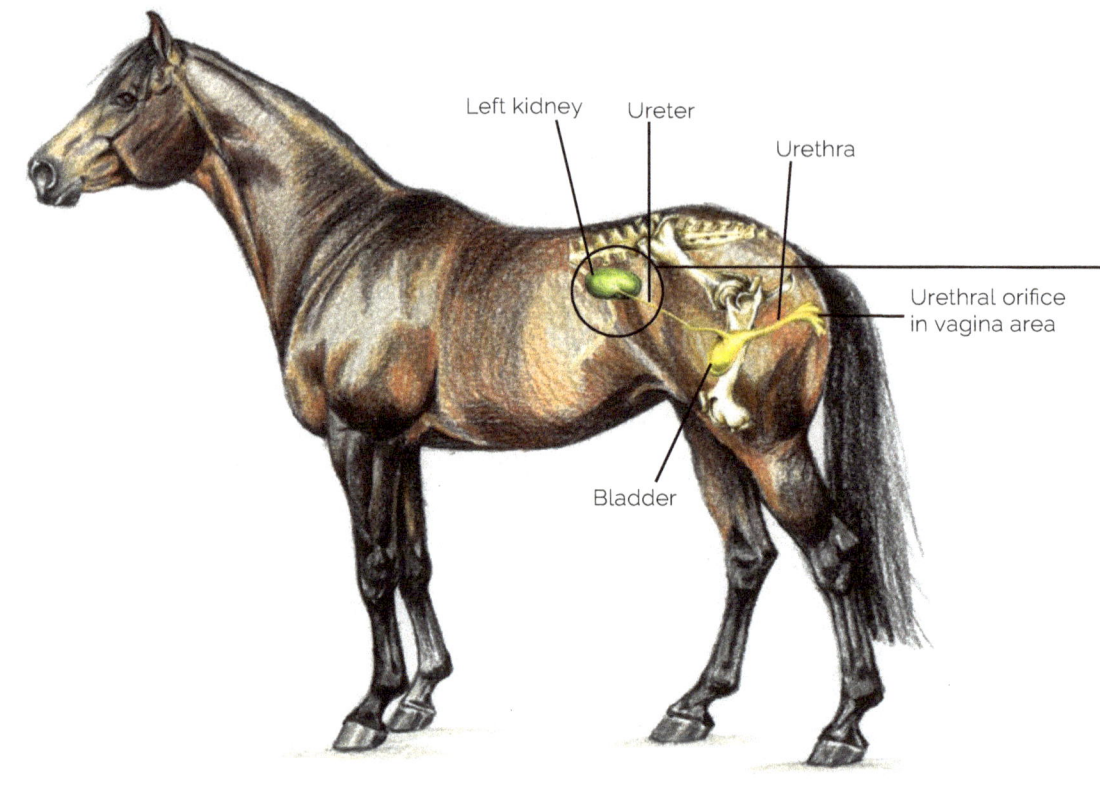

released with the bile, mainly leave the body via the intestines. The lungs are another disposal point where carbon dioxide and other gaseous substances are released. Most liquid waste products are excreted by the kidneys. The kidneys are located on the left and right of the abdominal cavity, below the spinal column, in the loins, i.e. at the transition between thoracic and lumbar vertebrae. In horses, the left kidney is bean-shaped and the right is more heart-shaped because it has to make space for the liver.

The liver converts waste products so that they can be excreted by the kidneys. These converted waste products are taken away by the blood circulation, which is how they get to the kidneys. Blood vessels enter each kidney on its concave side. They keep on branching out in the kidney and end in little balls made up of very fine capillaries. These capillary balls are surrounded by capsules made from renal tissue. Pressure in these capillaries is very high because of their small diameter and the large quantity of blood (around 25 percent of blood volume). The pressure squeezes the liquid out of the capillaries and into the surrounding capsule. This liquid, which is actually blood without cell constituents,

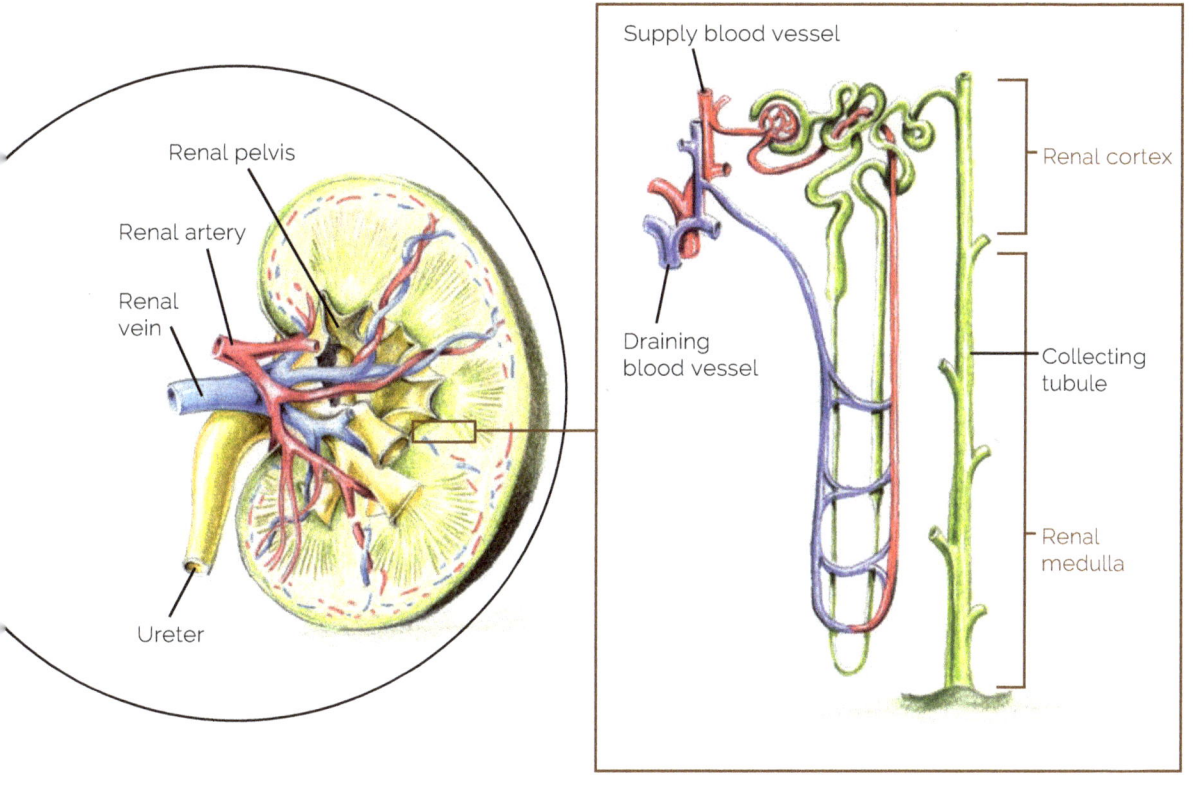

The urinary tract using the example of a mare, showing the structure of the kidney and the renal tubules where urine is produced. (Illustrations: Retsch-Amschler)

is called primary urine. Initially, it contains all dissolved substances that are also present in the blood, including ones that the body does not want to excrete. The quantities are also immense, with the horse producing around 550 litres of primary urine in 24 hours!

The role of the kidney is now to reduce the quantity and to reclaim water and other important nutrients. This naturally includes valuable blood sugars, but also minerals and trace elements, small proteins and other chemical compounds. Conversely, the kidney can also add to the urine specific waste products that are too large to pass through the capillary walls or that are present in large quantities, such as urea, the waste product of protein metabolism. Substances such as these are correspondingly "marked" by the liver, so that the kidney recognises them and adds them to the urine. The horse's urine is slightly base and very chalky, because horses absorb large quantities of calcium by eating grass and hay. It is therefore always a little milky. Following concentration in the renal tubules, around three to ten litres of secondary urine remain, which gets into the renal pelvis and then into the bladder via the urethra. The urine is collected here until it is excreted.

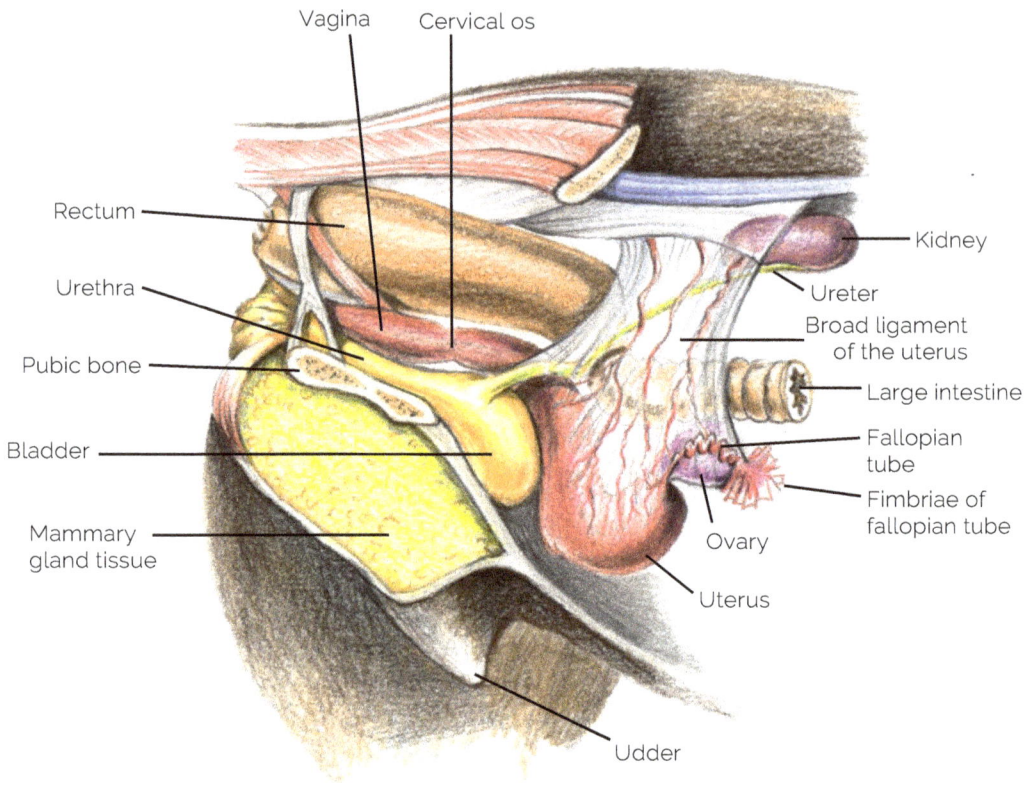

The sexual organs of a mare. (Illustration: Retsch-Amschler)

The bladder is fitted with a warning system that sends a message to the brain when it is full. The horse then adopts the typical rocking horse position in order to empty the bladder via the urethra. In mares, the urethra is very short and ends under the dock in the region of the vagina. In geldings and stallions, the urethra is much longer and joins up with the vas deferens from the testicles on the way. It ends at the tip of the penis. Specific voiding of the bladder is important so that urine is not running out constantly. Urine is very aggressive because it contains so many toxins and would irritate the skin. In addition, horses are normally very clean animals that do not urinate or defecate where they eat.

Sexual Organs and Reproduction

Mares and stallions have different reproductive organs, some of which can be seen from the outside, but the majority of which are inside the abdominal cavity or rather in the pelvic cavity. Mares reach sexual maturity at around 18 months old and, in herds in the wild, are covered by the stallion for the first time at around this age. However, because fillies are not fully mature at this age and because pregnancy is a very physically demanding experience, responsible breeders wait until the mare is three to four years old. Colts are sometimes sexually mature at nine months old and should be removed from mixed herds in good time.

A mare's cycle last approximately three to four weeks. During this time, a follicle ripens in one of the ovaries, which then releases the ripe egg. This egg travels through the fallopian tube in the direction of the womb. If the egg is fertilised it embeds into the womb and a embryo begins to grow. If the egg is not fertilised, it dissolves, the mucous membrane of the womb is built up again and the follicle ripening cycle starts from the beginning. The cycle is driven by hormones, but in horses it is not just controlled by the body's own sex hormones, which are activated by sunlight among other things, but also by beta carotene, large quantities of which are present in fresh pasture. That is why mares come into season, i.e. show their willingness to mate, as soon as the pasture season begins. "Mareish" behaviour decreases again in the winter.

Mares show their readiness to mate by constantly releasing small quantities of urine that is enriched with pheromones to attract stallions. As she does this, the mare carries her tail to the side and opens and closes her labia, which is called "winking". Stallions can tell when a mare is fertile by smelling her urine. He will show the typical flehmen response as he does this.

The flehmen response enables stallions to smell mares' pheromones particularly well.
(Photo: Slawik)

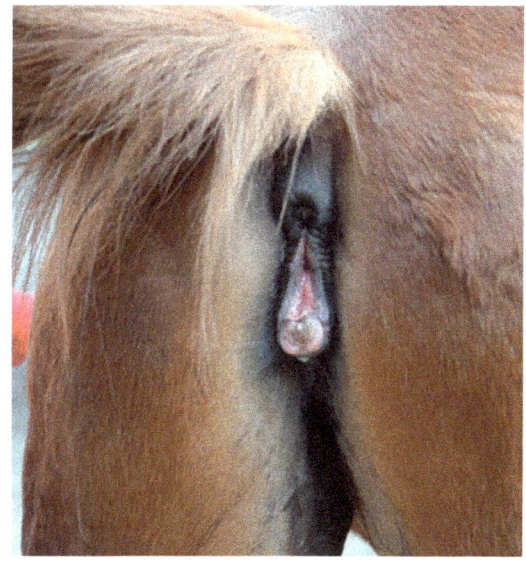

The mare "winks" to show her readiness to mate.
(Photo: Slawik)

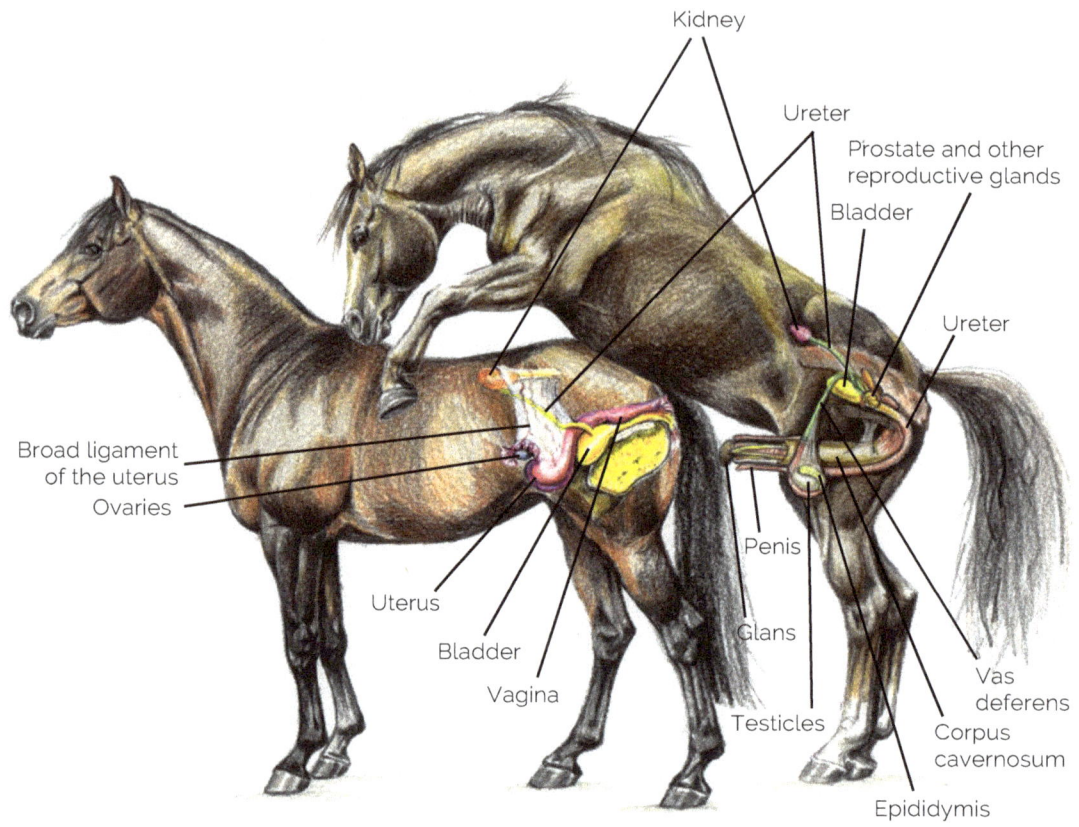

Mare and stallion during mating. (Illustration: Retsch-Amschler)

In stallions, spermatozoa mature in the testicles and epididymides, where they are stored until ejaculation. If there is a mare in season nearby, the stallion's penis will emerge from the sheath in which it is normally (apart from during urination) concealed. The penis becomes erect because of accumulation of blood. If the mare is ready to mate, the stallion may mount her. As he does so, he inserts his penis into the mare's vagina. When the stallion ejaculates, semen moves out of the epididymis along with secretions from the prostate and other reproductive

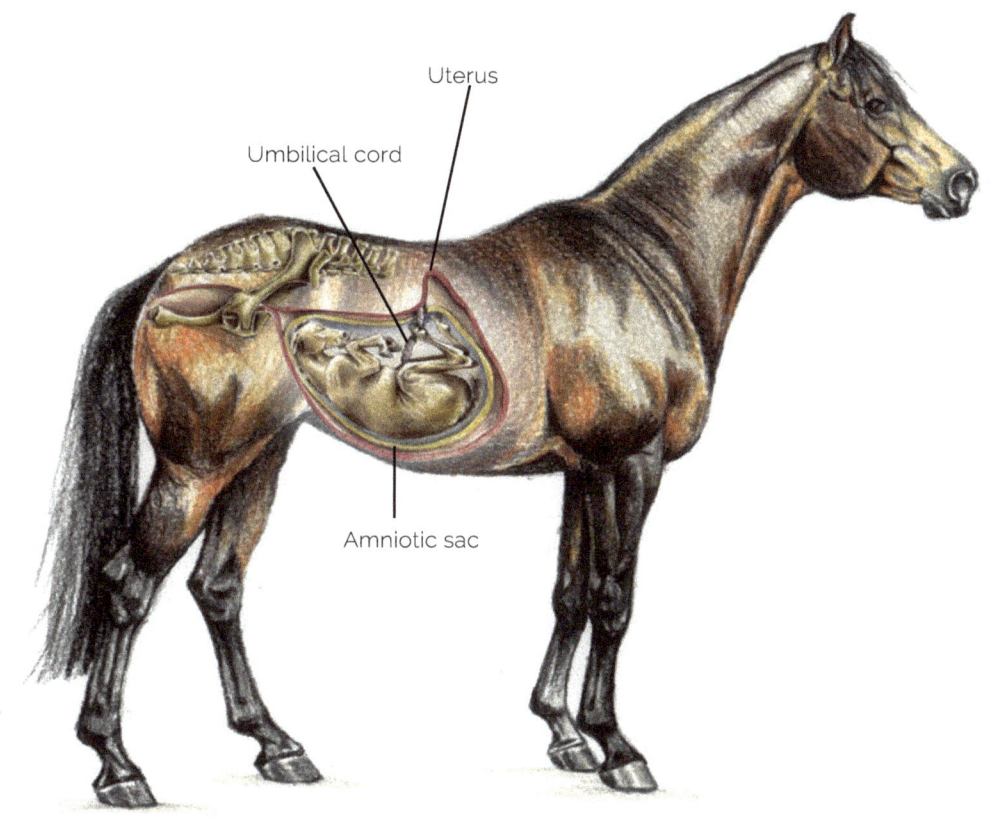

The position of an approximately 300 day-old foal in the womb, i.e. around 40 days before birth. The foal is now around 90 centimetres long and weighs 40 kilograms. (Illustration: Retsch-Amschler)

glands and is emptied into the vagina. The sperm start their journey through the uterine orifice in the direction of the fallopian tubes in order to fertilise the egg there.

The fertilised egg begins to divide in the fallopian tube and becomes a cluster of cells, a foetus. When it arrives in the womb (*uterus*), it implants into the mucous membrane, which is well supplied with blood, and forms an amniotic sac and a placenta and umbilical cord for receiving nutrition from the mare. The foetus begins to grow. The gestation period of a mare lasts for around eleven months. As it grows, the foal displaces its mother's intestines to make

Lots of turnout from the first day onwards is essential for healthy development of the foal. (Photo: Slawik)

space for itself. Like all mammals, foals are born head-first. When labour pains begin, mares usually lie down. The mare may also give birth standing up if the horses are kept in a herd where she is a low-ranking member. The foal's front hooves come first during the birth and they usually break through the amniotic sac. The head then appears between the front legs. Once the chest is through the birth canal, the rest of the foal usually slips out unaided. The mare then removes the amniotic sac and licks the foal. This process not only stimulates the foal's circulation, but creates the bond between mare and foal. If a person intervenes at this point to "help" the mare, the mare may reject the foal and it will have to be hand-reared.

As flight animals, foals can stand and walk shortly after birth. However, they are still unsteady on their legs to begin with and it can take a few days before they are frolicking around on their long foal legs. Foals are born with very steeply angled fetlock joints. The flexor tendons yield during the first two weeks after the birth and the actual fetlock angle appears. The softer the ground and the less the foal moves, the steeper the fetlock will remain. Lots of turnout from the first day on is important for healthy foal development.

Pigmented cells (melanocytes) are responsible for dark pigmentation of the skin and coat. Cells in the pale areas of skin do not contain any melanin, so the blood circulation makes them appear pink. (Photo: Slawik)

The Skin and Skin Appendages

The skin *(cutis)* is also called the "mirror of health". Skin is not just a mechanical protection for all of the underlying structures and a part of the immune system. The skin is also a excretive organ, it is involved in thermoregulation, it is a sense organ for touch and pain, a store for fat and electrolytes and, last but not least, it forms protective appendages such as fur and hooves.

The skin is made up of three layers. The top layer *(epidermis)* is primarily a mechanical protective layer. It protects against injury, pathogens, heat and cold.

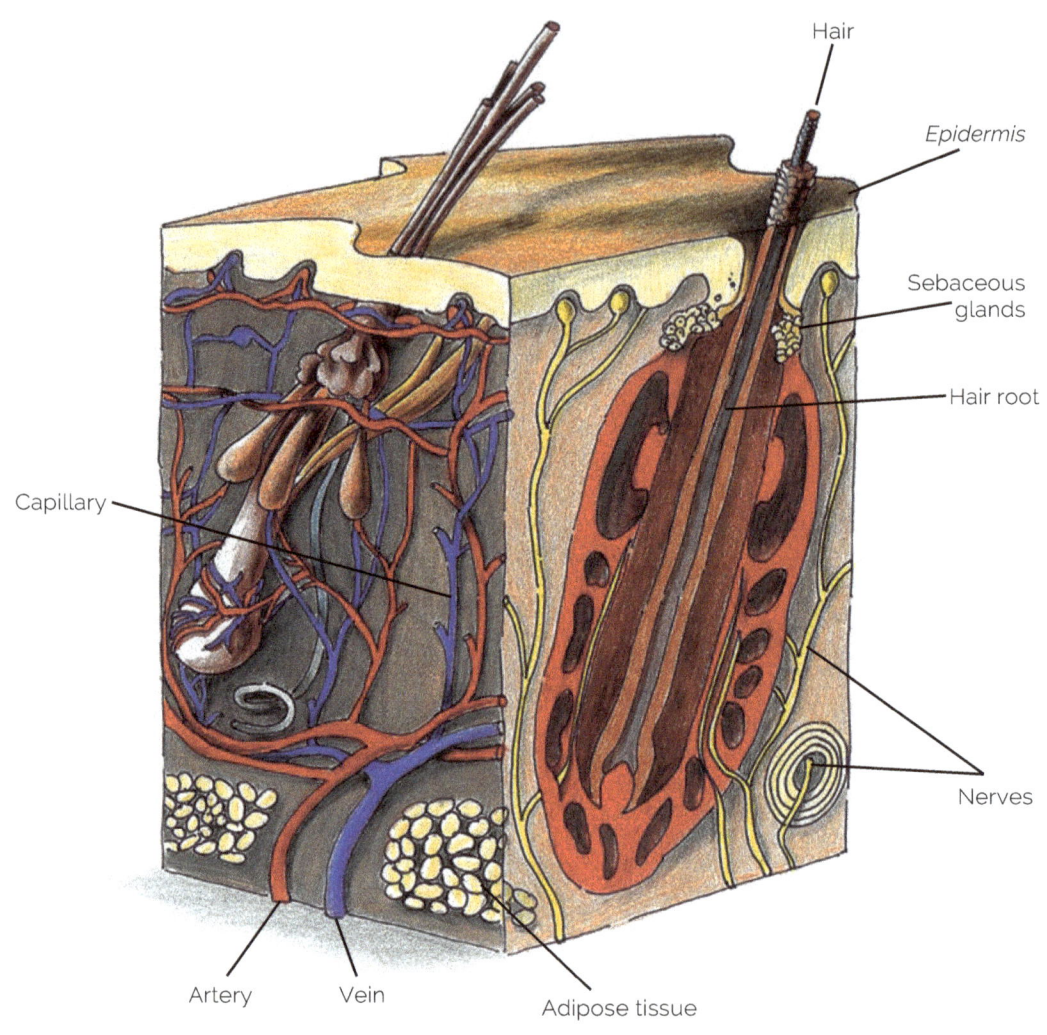

The skin is a complex organ with many important functions. (Illustration: Retsch-Amschler)

Depending on its use, the epidermis can become horny, as we can see in the hooves or sometimes in old rug pressure points on the withers. It has a coating of endogenous fatty acids that come from the sebaceous glands and perspiration and form the skin's "acidic protective coat". The grease is water-repellent and keeps the skin and coat smooth. It is also an assembly area for skin flora, symbiotic bacteria that colonise the skin and provide the horse with extra protection against germs. The melanocytes are found in the lower layers of the epidermis. They are pigmented cells that give the skin its dark colour and the coat and/or mane and tail their colour.

Did you know ...?

Washing with shampoo destroys the horse's natural acidic, protective coat. The skin becomes dry and starts to itch and rubbing causes tiny tears in the skin where mites, fungi and other parasites can become established. If you have to wash your horse you should only use water. If you are using cold water, always start by spraying the water at the hooves and move slowly in the direction of the heart so that the circulation does not suffer a shock.

The *corium*, the layer of skin used to make leather, lies beneath the epidermis. The corium contains cutaneous systems such as hair roots, nerve receptors for pressure, pain or heat, as well as sebaceous glands and sweat glands and the *arrectores pilorum muscles*, which enable the hairs to stand on end. The sweat glands work in a similar way to the kidneys and they can excrete waste products if the kidneys are overloaded. That is why horses smell unpleasant when they have metabolic disorders. Along with salts, horse sweat also contains many proteins, which is why it foams when rubbed.

Numerous vessels that are involved in thermoregulation run through the corium. They can be widened if it is very warm to allow the blood to be cooled on the surface of the body. If the horse is too hot, sweat also cools the skin through evaporation. In heat and during physical work, horses can produce up to 10 litres of sweat per hour, losing 50 grams of sodium at the same time. The blood vessels of the corium narrow under cold conditions so that not too much body heat is lost through the skin. The blood vessels in the skin also narrow during shock, so horses will usually feel cold after an accident.

The skin also contains many white blood cells or leucocytes. They are on standby should an injury and penetration by pathogens occur.

Did you know ...?

The horse's skin and coat are a thermal miracle. When it is cold, the arrectores pilorum muscles make the hairs stand on end, like goosebumps in people. The air becomes trapped between the hairs as a result, heats up and creates an insulating cushion of warmth. As well as the dense undercoat, winter coats also have hollow hairs with an additional insulating effect. A horse with a normal winter coat therefore has a down jacket and a thin shirt in one, depending on whether the hairs are erect or flat. The horse's coat also acts like a sweat rug. Sweat is rapidly transported to the outside so that skin dries off quickly again after sweating, even when the surface of the coat is still damp. Therefore, in winter, sweating is often healthier than putting unsuitable or dirty thermal blankets on a clipped horse.

The nervous system controls all voluntary and involuntary functions in the body, including the startle and fight or flight reflexes. (Photo: Slawik)

The Nervous System

The nervous system is made up of two types of cell, nerve cells *(neurones)* and neuroglia. Nerve cells have a typical shape with a cell body and two types of processes: short dendrites, through which signals come into the cell and long axons, through which signals are passed on to other cells. Signals are transferred as electrical impulses within nerve cells. Synapses are located between two nerve cells or at the end of a signal pathway. Signal transmission takes places chemically at the synapses, with the help of transmitters. The signal can only ever be passed on in one direction at a synapse. This ensures that the pathway cannot suddenly change direction, but that every nerve fibre is a one-way street.

So that electrical transmission of impulses within the cell is as fast as possible, the axon is covered with a kind of insulating layer. This insulation is formed by neuroglia and is also called the myelin sheath. It is interrupted at regular intervals by myelin sheath gaps. This construction makes the transfer faster because the impulse "jumps" from gap to gap. Together with the hormone system, the nervous system controls all of the processes in the body, whether they are voluntary or involuntary. With regard to the nervous system, we differentiate between the central nervous system (CNS), the autonomic nervous system (ANS) and the peripheral nervous system (PNS), according to the functions performed.

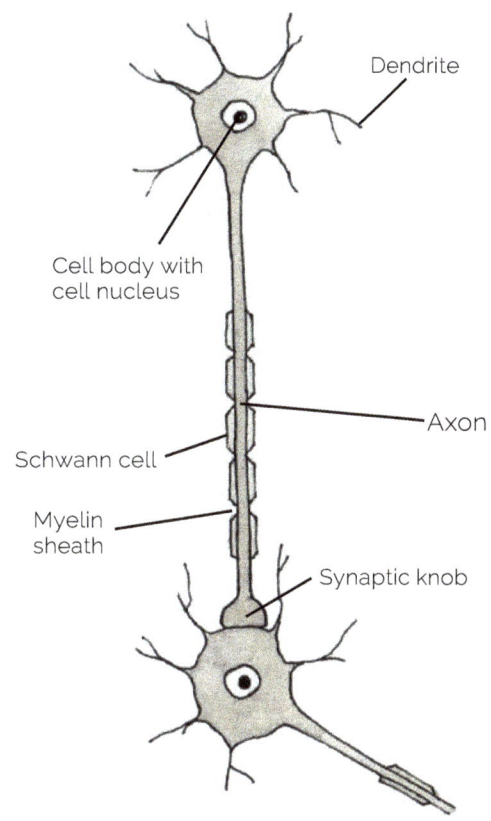

The nerve cell has a cell body and two types of processes. (Illustration: Mähler)

The Central Nervous System

The central nervous system (CNS) comprises the brain inside the skull and the spinal column that runs through the spinal canal, which is formed by the individual vertebrae. Exit channels are found between each of the vertebrae where nerves leave the spinal column and go into the body (peripheral nervous system). Signals from the body also reach the spinal column through these openings and are forwarded on to the brain from there. The spinal column ends in the area of the lumbar spinal column and only nerve strands continue from this point to emerge at the sacrum at the furthest point.

Did you know …?

We can picture the nervous system in the body as a bundle of power lines that come from the power station (brain). Along the spinal cord, bundles of cables branch off out of this bundle and go to the different areas (for example, a forelimb or hind limb). They branch off again within each area into the different streets (muscles) until they finally arrive at the house (muscle fibre) as a single strand. The spinal cord is the large bundle of cables that is part of the central nervous system. The branches represent the peripheral nervous system.

As an "extension" of the spinal column, the brain is responsible for the most diverse range of conscious and unconscious bodily processes. (Illustration: Retsch-Amschler)

The brain is actually an extension of the spinal column that performs various tasks. This is where conscious control of the body takes place, i.e. thought and learning processes (cognition), but also monitoring and control of all unconscious bodily processes such as breathing, intestinal motility or reflexes. Information from the body, about the tonicity of a muscle for example, is forwarded from the peripheral nervous system, via the spinal cord, to the brain. The brain compares the tonicity with the normal tonicity stored and with the other muscles and then sends signals via the spinal cord into the peripheral nervous system, which makes sure that muscle tonicity is regulated.

The brain has different regions. The cerebrum is mainly responsible for thought processes, for conscious processing of information from the sense organs and for conscious control of movements. The cerebellum, which looks like a cauliflower and is located between the cerebrum and the spinal column, is also involved in movement. It translates the commands that come from the cerebrum into coordinated muscle actions.

The diencephalon occupies a central position in the brain. This is where most information from the periphery and the sense organs initially arrives and where meaning is attributed to this information.

Important information is forwarded on to the cerebrum and unimportant information is discarded. Information about danger ensures that the fight or flight reflex is activated immediately, without asking the cerebrum first. These reflexes are controlled by the midbrain (mesencephalon), which is directly underneath the diencephalon. Most reflex centres are found in the midbrain, not just for flight, fight and defence, but also for the degree of alertness and sleepiness.

The Autonomic Nervous System

The diencephalon and midbrain work closely together with the autonomic nervous system (ANS). The ANS has two antagonists: the sympathetic nervous system and the parasympathetic nervous system. They control all involuntary functions, from intestinal motoricity to heart rate. In order to do this, they work closely together with the hormone system that regulates subconscious processes. The sympathetic nervous system runs parallel to the left and right of the spinal column ("sympathetic chain") and is responsible for all activating processes and those that use up energy. The sympathetic nervous system accelerates the heart rate, reduces circulation and motoricity in the digestive tract, narrows the blood vessels in the skin and in the liver and diverts the blood into the skeletal muscles. It becomes active when the horse is preparing itself for fight or flight.

The parasympathetic nervous system is the antagonist of the sympathetic nervous system and it takes care of all body

The parasympathetic nervous system is responsible for all bodily reactions that have to do with rest and energy yield. (Photo: Slawik)

reactions that allow the horse to come to rest and that are used for energy yield. For example, it slows down the heart rate, increases circulation in the intestines and liver and stimulates intestinal motoricity. The parasympathetic nervous system features various nerve strands that originate directly from the brain, in particular the tenth cranial nerve *(nervus vagus)* that innervates all of the intestines. Furthermore, the PNS also includes nerves in the rear lumbar area that come from the spinal column and go to the bladder, the rectum and to the sexual organs.

The Peripheral Nervous System

The peripheral nervous system includes all nerves that run in the body and that do not belong to the CNS or the ANS. We can differentiate between motoric and sensory nerve fibres. Motoric nerve fibres come from the spinal column and innervate the muscle cells via the motor end-plate. The sensory nerve fibres come from receptors in the muscles, the skin and all other sense organs and run from them in the direction of the spinal column. In comparison with the other nervous systems, the peripheral nervous system is special because these nerves can regenerate when they are injured. While injuries to the brain or spinal column are irreparable and can lead to severe ataxia or paralyses in horses, peripheral nerves can regenerate with time and can grow together again. This can even happen to operatively separated nerves after a few years have elapsed.

The Horse's Senses

Horses have sense organs so that they can perceive their environment and react to it. The senses include: seeing (eyes), hearing (ears), smell (nose), taste (tongue), balance (labyrinthine system in the ear) and touch, which includes the tactile hairs on the muzzle as well as pressure, pain and temperature receptors. As you can see, most senses are found on the head itself. This allows stimuli to be quickly forwarded to the brain and information to be processed immediately.

The eye is one of the horse's essential senses, which it uses to perceive visual stimuli. As with all flight animals, the horse's eyes are located on the side of its skull in order to give a 360 degree view, if possible. The horse's only blind spots are immediately in front of and behind it. Furthermore, the pupil is a horizontal bar that gives a larger, horizontal picture so that the horse can observe its whole environment without having to turn its head. That is also why horses' eyes cannot focus particularly well. Instead, the eye has a similar structure to a varifocal lens. When the head is lowered, the horse sees close-up things clearly, i.e. what is in front of its feet and what it is eating. When the head is raised, the horse can see distance clearly and can perceive threats on the horizon. That is why horses raise their heads and scan their surroundings when they get a fright and why, conversely, they lower their head when there is something worrying right in front of them.

This practical set-up comes at the expense of spatial vision and sharp focus. This can be best observed in young show jumpers that lower their heads just before the jump in order to precisely home in on the

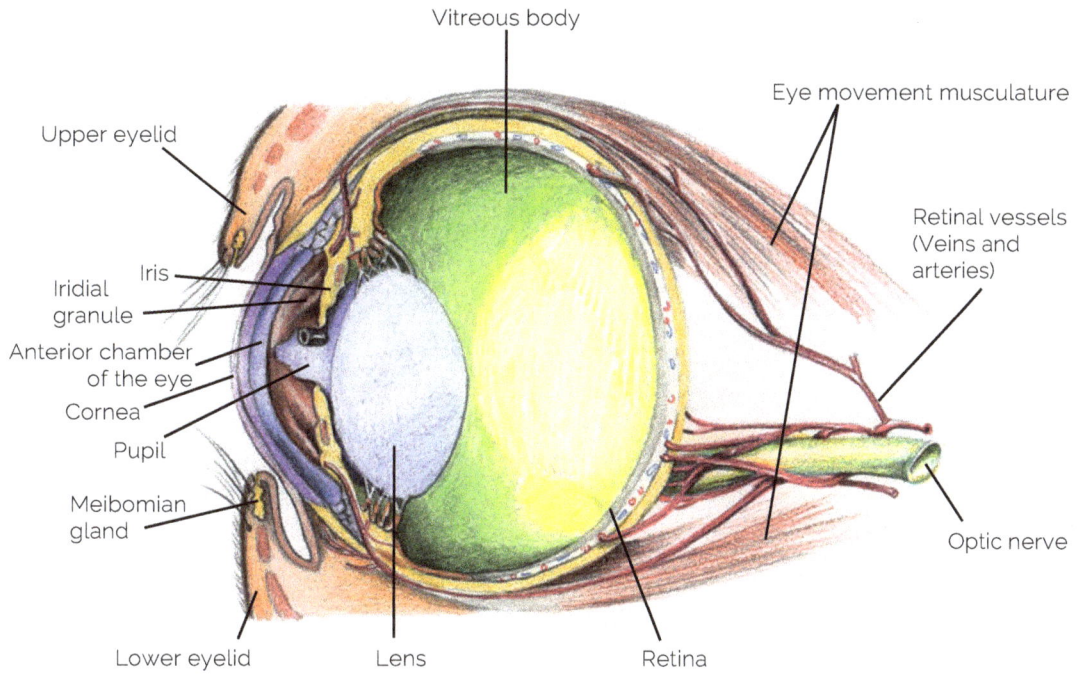

Structure of the equine eye. (Illustration: Retsch-Amschler)

obstacle before take-off. Horses also perceive colours in a slightly different way from humans. This puts them in an excellent position for seeing at dusk and they are much better than us at seeing movements in front of a still background. Light that comes through the pupils and falls onto the retina stimulates receptors that forward their information via the optic nerve to the visual cortex of the brain. The eye needs to be constantly wetted by tear fluid to stop it from drying out.

The horse's sense of smell is in its nose or, more precisely, in the ethmoid bone. The mucous membrane of the ethmoid bone contains various receptors for a wide range of odourant

The horse's field of vision. The horse can see best in the green area. It cannot see the white area clearly and objects within the red area are in the blind spot. (Illustration: Mähler)

The Nervous System

Horses use smell to recognise members of their herd. (Photo: Slawik)

molecules that are activated as soon as a corresponding odourant molecule binds with them. This information is then passed onto the rhinencephalon where it is processed. Smell is one of the most primitive senses, much older than vision. In a similar way to dogs, horses use their sense of smell much more intensely than we humans. They use it not only to recognise edible food, but also members of their herd, their environment and much more.

The sense of hearing is located in the inner ear. Sound meets the auricle or external ear. In horses, the opening of the auricle is protected against dirt and insects by bushy hair. Horses are able to swivel their ears in all directions in order to precisely locate where a sound is coming from. The cartilage in the auricle breaks up the sound so that it meets the eardrum perfectly, which then begins to vibrate. It passes on its vibration to three small bones, the hammer, anvil and stirrup bones. These are suspended in the middle ear by tiny muscles. Contraction or relaxation of these muscles can regulate the extent to which vibrations of the eardrum are forwarded onto the inner ear. In a very loud environment, all sounds are muffled when they are passed on so as not to damage the ear. Conversely, these three small bones can also amplify very quiet sounds. Finally, the cochlea is located in the inner ear. Its receptors pass on the sound stimuli into the auditory centre of the brain. This is also the location of the labyrinthine system, which gives the brain information about the movement of the body in space.

The horse's sense of touch is much more sensitive than that of a human being. For example, it can very accurately perceive light skin contact, such as a fly landing. There are also hairs (*vibrissae*) on the head around

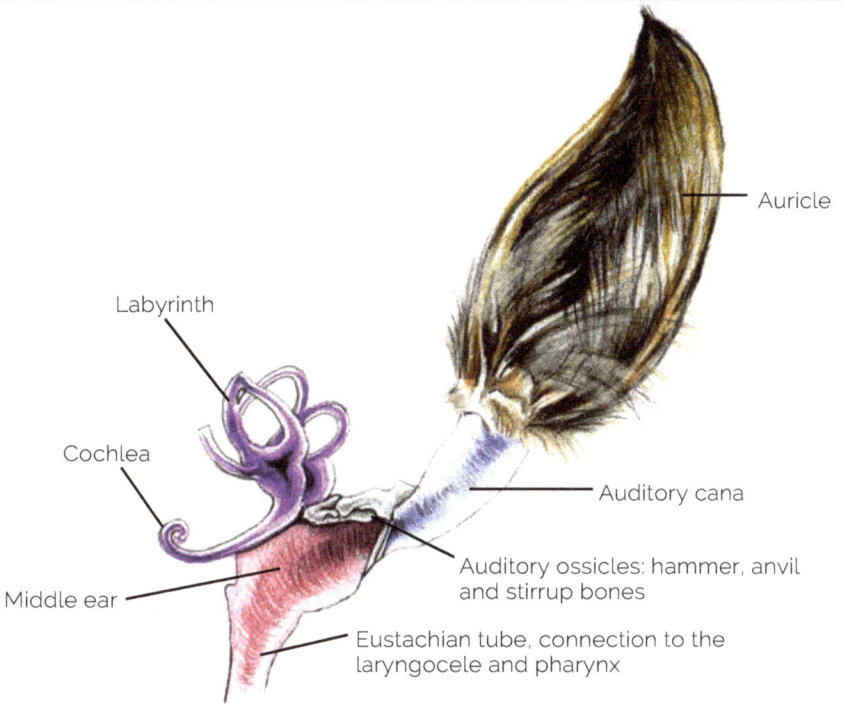

the muzzle and eyes with receptors at their roots that send information about movement to the brain. These hairs are often removed by clipping, which robs the horse of an important sense organ.

The sense of taste is also important for horses so that they can differentiate between edible and unpalatable or poisonous plants. This knowledge is not innate in horses, but is something that they have to learn. If a horse eats a plant and feels ill afterwards, he will avoid plants that taste like that in the future. Without a sense of taste, a horse would eat poisonous plants in pasture that others would leave. Taste buds on the tongue act as receptors for different tastes such as sweet, sour, salty and bitter. In order to differentiate more accurately between flavours, the sense of smell is also involved.

The horse's organs for hearing and rotation/balance are in the inner ear and are connected to the auricle, which is visible from the outside, by the middle ear. (Illustration: Retsch-Amschler)

The hairs on the muzzle are one of the horse's essential sense organs and they should not fall victim to clippers. (Photo: Slawik)

The Nervous System

Index

The Skeleton
2nd phalanx/short pastern (*phalanx media*)...38 ff
Articular cartilage .... 15 ff
Ball-and-socket joint 18, 19, 31, 35
Bicondylar joint 20
Bone cells (*osteocytes*) .... 13 ff
Bone marrow 15
Cannon bone (*os metacarpale III*) 23, 30, 39, 40
Carpal joint/knee joint (*articulatio manus*).....8, 9, 15, 18, 19, 31, 32, 60
Carriage joint 20, 37
Cartilaginous joint 22, 31, 37
Caudal spinal column 25
Cervical vertebrae .... 19, 22 ff
Cochlear joint .... 20 f
Coffin joint (*articulatio interphalangea distalis*)....38 f
Common digital extensor tendon 40
Condyloid joint 19, 24
Corium .... 38 f, 85
Cortical/compact bone 14
Corticalis 14
Digital annular ligament .... 40 f
Digital cushion 38
Elbow joint (*articulatio cubiti*) 20, 31, 48,
Epiphysis 15
Ethmoid bone .... 26, 66 f, 91
Exostosis .... 13 ff
Facet joints 26
Femur (os femoris) 21, 35, 37, 47, 54, 55
Fetlock (*articulatio metacarpophalangea*)......9, 39, 40 f
Flexor tendons .... 31, 37 ff, 82
Forearm (*os radius*) 8, 31, 48
Frontal bone 16, 26, 27
Gaskin (*os tibia*) .... 37, 53 ff
Hinge joint 20, 31, 39
Hip joint (*articulatio coxae*) 19, 33, 35, 47, 54
Hock joint (*articulatio tarsocruralis*) 8, 10, 20, 32, 37, 52, 54, 55, 60
Hoof capsule .... 38 f
Hoof wall 38
Incisive bone (*os incisivum*) 27
Joint capsule... 15, 17 f
Lacrymal bone (*os lacrimale*) 26
Long pastern/
1st phalanx (*phalanx proximalis*) .... 39, 40 f
Lower jaw (*os mandibulare*) 59
Lumbar vertebrae .... 24 f, 47, 76, 87
Mandibular joint 18f, 26
Medullary cavity .... 14 f
Meniscus .... 18 f, 36, 37
Nasel bone 26
Navicular bone .... 38 f
Nuchal crest 26, 27
Occipital bone (*os occipitale*) .... 26 f

Os tympanicum 26
Pastern joint
(*articulatio interphalangea proximalis*).........28 ff, 54
Pedal bone (*phalanx distalis*) .... 38 ff
Pelvis (*os pelvis*) .... 10, 16 , 22, 25, 33, 34, 54,
Periosteum 13ff
Pivot joint 21, 24
Plane joint 19
Podotrochlear apparatus 38
Pubic symphysis 16
Ribs .... 14, 25, 68 f
Sacroiliac joint... 22, 25, 33, 34
Sacrum .... 9 f, 22, 25 f, 33, 34, 47, 54, 87
Saddle joint 20, 39
Shoulder blade (*os scapula*) 19, 30, 31, 46, 48, 50, 51
Shoulder joint (*articulatio humeri*) 31
Skull .... 16, 19, 26 f, 66, 87
Sole 38, 41
Spinal canal .... 25 f, 87, 90
Spinous process 25
Spiral joint .... 20 f
Spongiosa 14
Sternum 24, 25, 61
Stifle joint .... 18 f 55
Suture 16, 26
Synovial fluid .... 17 f
Teeth .... 27 ff, 69
Temporal bones (*Ossa temporale*) 16, 26
Tendon sheath 41, 43
Thoracic vertebrae 25
Thorax 24f, 30, 61, 63, 68
Trabeculae .... 36 f
Upper arm (*os humerus*) 19, 31, 46, 48, 50
Upper jaw (*os maxillare*) 26, 27
White line 38
Withers 8, 25
Zygomatic arch (*arcus zygomaticus*) 26, 27
Zygomatic bone (*os zygomaticum*) 26

The Muscles
Bucksaw construction 60, 63, 89
Cardiac musculature .... 43 f
Fascia 30, 34
M. abductor pollicis longus.................................. 50
M. biceps femoris 54
M. brachiocephalicus 46, 50
M. deltoideus 50
M. extensor digitorum lateralis............................ 54
M. extensor digitorum longus 54
M. fibularis tertius 54
M. flexor digitorum profundus 54
M. flexor digitorum superficialis 54
M. gastrocnemius 54
M. glutaeus accessorius 54
M. glutaeus medius 54
M. glutaeus profundus 54
M. glutaeus superficialis....................................... 54

M. gracilis	54
M. iliocostalis	47
M. iliopsoas	47
M. infraspinatus	50
M. latissimus dorsi	50
M. longissimus	47
M. longissimus atlantis	46
M. longissimus capitis	46
M. obliquus externus abdominis	47
M. obliquus internus abdominis	47
M. omotransversarius	46, 50
M. pectineus	54
M. pectoralis ascendens (profundus)	49, 50
M. pectoralis descendens	50
M. popliteus	54
M. quadrizeps femoris	54
M. rectus abdominis	47
M. rhomboideus	46, 50
M. sartorius	54
M. scalenus	46
M. semimembranosus	54
M. semitendinosus	54
M. serratus ventralis cervicis	46, 49, 50
M. serratus ventralis thoracis	49, 50
M. spinalis	47
M. spinalis cervicis	46
M. splenius	46
M. sternocephalicus	46
M. subclavius	49, 50, 54
M. subscapularis	50
M. tensor fascia latae	54
M. tibialis caudalis	54
M. tibialis cranialis	54
M. transversus abdominis	47
M. trapezius	46, 50
M. trapezius cervicis	50
M. trapezius thoracis	50
Mm. adductores	54
Mm. multifidii	47
Muscle fibres	42, 44, 87
Muscle fibrils	42
Sarcomeres	42
Skeletal musculature	42 ff
Smooth musculature	43 f, 72

The Organs

Alveoli	60 f, 67 f
Aorta	58
Arteries	58 ff
Bladder	75 ff
Blood	41, 56 ff, 72, 74, 85, 89
Bronchi	67f
Caecum	72 f
Colon	73
Diaphragm	67 f, 70, 74
Duodenum	72
Ear	90, 92
Epididymis	79 f
Eye	26, 90, 91
Frog	41, 60
Gastric acid	71
Heart	25, 41, 56 f, 61 ff
Ileum	72
Jejunum	72
Kidney	74, 75 ff
Large intestine	72 ff
Larynx	67, 70
Leucocytes	57 f, 62
Liver	58, 65, 72, 74, 76 f
Lung	25, 57, 60 f, 63, 66 ff, 76
Lymph	60, 64 f
Lymph nodes	64 f
Nose	66, 90 f
Oesophagus	67, 70
Ovaries	78
Pancreas	72, 74
Penis	79 f
Plasma	57, 60
Pulmonary circulation	56, 60
Rectum	74, 90
Salivary glands	70
Serum	57
Skin	65, 83 ff, 92
Small intestine	71 ff
Spleen	57
Stomach	70 ff
Systemic circulation	56, 58, 60, 63
Testicles	78 ff
Thrombocytes	57 f
Tongue	69 f, 90, 93
Trachea/windpipe	67 f, 70
Uterus	78 ff
Vagina	80
Veins	58 ff
Vibrissae/whiskers	90, 92 f

The Nervous system

Autonomic nervous system	63, 87, 89
Axon	86 f
Brain	44, 87 f, 90 ff
Central nervous system	87
Dendrit	86 f
Motor end-plate	44, 90
Myelin sheath	87
Myelin sheath gaps	87
Nerve cells (neurones)	58, 86
Neuroglia	86 f
Parasympathetic nervous system	89
Peripheral nervous system	87 f, 90
Spinal column	26, 87 f, 90
Sympathetic nervous system	89
Synapses	86

Xenophon Press Library

www.XenophonPress.com
Xenophon Press is dedicated to the preservation
of classical equestrian literature.
We bring both new and old works to English-speaking riders.

30 Years with Master Nuno Oliveira, Henriquet 2011

A Journey Through the Horse's Body, Fritz 2012

A Rider's Survival from Tyranny, de Kunffy 2012

Another Horsemanship, Racinet 1994

Austrian Art of Riding, Poscharnigg 2015

Broken or Beautiful: The Struggle of Modern Dressage, Barbier/Conrod 2020

Classic Show Jumping: the de Nemethy Method, de Nemethy 2016

Classical Dressage with Anja Beran, Beran 2017

Divide and Conquer Book 1, Lemaire de Ruffieu 2016

Divide and Conquer Book 2, Lemaire de Ruffieu 2017

Dressage for the 21st Century, Belasik 2001

Dressage in the French Tradition, Diogo de Bragança 2011

Dressage Principles and Techniques: A Blueprint for the Serious Rider, Tavora 2018

Dressage Principles Illuminated, Expanded Edition, de Kunffy 2021

École de Cavalerie Part II, Robichon de la Guérinière 2015

Elements of Dressage, von Ziegner 2016

Equestrian Art: The Collected Later Works, Nuno Oliveira 2022

Equestrian Art: The Collected Early Writings (1951-1956), Nuno Oliveira 2022.

Equine Osteopathy: What the Horses Have Told Me, Giniaux 2014

Equitation, Bussigny 2021

Federico Grisone's "The Rules of Riding," Grisone/Tobey 2022

Fragments from the Writings of Max Ritter von Weyrother, Fane 2017

François Baucher: The Man and His Method, Baucher/Nelson 2013

General Chamberlin: America's Equestrian Genius, Matha 2020

Great Horsewomen of the 19th Century in the Circus, Nelson 2015

Gymnastic Exercises for Horses Volume II, Eleanor Russell 2013

H. Dv. 12 German Cavalry Manual of Horsemanship, Reinhold 2014

Handbook of Jumping Essentials, Lemaire de Ruffieu 2015

Handbook of Riding Essentials, Lemaire de Ruffieu 2015

Healing Hands, Giniaux, DVM 1998

Horse Training: Outdoors and High School, Beudant 2014

I, Siglavy, Asay 2018

Horsemanship & Horsemastership Volume 1, US Cavalry 2021

Horsemanship Training Films 3 DVD set, US Cavalry 2021

Learning to Ride, Santini 2016

Legacy of Master Nuno Oliveira, Millham 2013

Lessons in Lightness: Expanded Edition, Mark Russell 2019

Mark of Clover, Kelly, 2022

Methodical Dressage of the Riding Horse, Faverot de Kerbrech 2010

Military Equitation or, A Method of Breaking Horses, and Teaching Soldiers to Ride, Pembroke, and *A Treatise on Military Equitation*, Tyndale 2018

My Horses Have Something to Say, de Wispelaere 2021

Principles of Dressage and Equitation, a.k.a. Breaking and Riding, Fillis 2017

Racinet Explains Baucher, Racinet 1997

Releasing the Jaw, Poll, and Neck DVD, Mark Russell 2021

Riding and Schooling Horses, Chamberlin 2020

Riding by Torchlight, Cord 2019

Riding in Rhyme, Davies 2021

Schooling Exercises In Hand, Hilberger 2009

Science and Art of Riding in Lightness, Stodulka 2015

Sketches of the Equestrian Art, Barbier/Sauvat 2022

The Art of Riding a Horse, D'Eisenberg 2015

The Art of Traditional Dressage, Volume 1 DVD, de Kunffy 2013

The Chamberlin Reader, Chamberlin/Matha, 2020

The de Nemethy Method: A training seminar, 8 DVD set, de Nemethy 2019

The Ethics and Passions of Dressage Expanded Edition, de Kunffy 2013

The Forward Impulse, Santini 2016

The Gymnasium of the Horse, Steinbrecht 2018

The Horses, a novel, Walker 2015

The Italian Tradition of Equestrian Art, Tomassini 2014

The Maneige Royal, de Pluvinel 2010, 2015

The New Method of Dressing Horses a.k.a. A General System of Horsemanship, Cavendish 2020

The Portuguese School of Equestrian Art, de Oliveira/da Costa 2012

The Quest for Lightness in Equitation and Equestrian Questions, Nelson/L'Hotte 2021

The Spanish Riding School & Piaffe and Passage, Decarpentry 2013

The Spanish Riding School: The Miracle of the White Horse DVD,
  US Lipizzan Association 2021

To Amaze the People with Pleasure and Delight, Walker 2015

Total Horsemanship, Racinet 1999

Training Hunters, Jumpers, and Hacks, Chamberlin 2019

Training Your Foal, Ettl 2011

Training with Master Nuno Oliveira, 2 DVD set, Eleanor Russell 2016

Truth in the Teaching of Master Nuno Oliveira, Eleanor Russell 2015

Wisdom of Master Nuno Oliveira, de Coux 2012

www.ingramcontent.com/pod-product-compliance
Lightning Source LLC
Chambersburg PA
CBHW041235240426
43673CB00011B/351